MODEL "M" MOCK UP
TOUCH CONTROL LIFT
M° 16028
2-5-44

I-49239

100 YEARS

FARMALL

FARMALL

100 YEARS

RANDY LEFFINGWELL

WITH ROBERT N. PRIPPS

INTERNATIONAL
TRACTOR WORKS
CHICAGO
ILL.
HARVESTER CORP.
INTERNATIONAL
INTERNATIONAL
8-16
KEROSENE
TRACTOR
INTERNATIONAL
8-16
KEROSENE
TRACTOR

CONTENTS

ACKNOWLEDGMENTS

As with my first tractor book in 1999, I must thank my friend and colleague Guy Fay. He has practically made the study of International Harvester Corporation (ICH), and its tractor and implement development, his life's work. His insight and understanding of how IHC and Farmall have affected the history of mechanized farming are second to none.

I wish to thank John Harper at CNH America LLC for access to information and photographs of the new Farmall model D and DX tractors. I am grateful to Jeff Walsh, former CNH Director of Communications, for making available QC-503.

Tractors inspire loyal legions of collectors. Among the most loyal are the enthusiasts I met and worked with while researching and photographing tractors for this book. A number of people opened their barns and sheds to me, washing up and pulling out a grand array of International Harvester's Farmall history.

I am deeply grateful to John and Jane Alling, Valley Center, California; Mike, Linda, and Eric Androvich, Grand Rapids, Ohio; Arden Baseman, Mosinee, Wisconsin; Dave and Anita Boomgarden, Chatsworth, Illinois; Vercel and Marilyn Bovee, Alto, Michigan; David and Ash Bradford, Warren, Indiana; Nate Byerly, Noblesville, Indiana; David and Gail Fay, Greenville, Pennsylvania; Keith and Cheri Feldman, Alto, Michigan; Lyman and Vivian Feldman, Alto, Michigan; Frank Ferguson, Decker, Michigan; Bob and Michelle Findling, Gladwin, Michigan; Allan and Joan Fredrickson, Lakewood, California; David and Carol Garber, Goshen, Indiana; Jack and Tammy Gaston, Athens, Ohio; Wilson and Portia Gatewood, Noblesville, Indiana; Jay Graber, Parker, South Dakota; David and Linda Grandy, Waconia, Minnesota; Becky and Rod Groenewald, Director, Antique Gas and Steam Engine Museum, Vista, California; Dave Hinds, Marion, Indiana; Joan Hollenitsch, Garden Grove, California; Ken Holmstrom, Harris, Minnesota; Wayne and Betty Hutton, Clarence, Missouri; Matt Jackson, Noblesville, Indiana; Kenny and Charlene Kass, Dunkerton, Iowa; Wendell and Mary Kelch, Bethel, Ohio; Jeff Kelich, Arcadia, Indiana; Tom and Mark McKinney, Noblesville, Indiana; Harold McTaggart, Port Hope, Michigan; Judy Meyer-Diercks, Stonefield Village, Wisconsin State Historical Society, Cassville, Wisconsin; Jerry and Joyce Mez, Avoca, Iowa; George and Barbara Morrison, Gladwin, Michigan; Ron Neese, Noblesville, Indiana; Robert Off, Tipton, Indiana; Scott Parsons, Oceanside, California; Jay Peper, Toledo, Ohio; Loren and Elaine Peterson, Sparta, Michigan; Bob and Mary Pollock, Dennison, Iowa; Fred and Janet Schenkel, Dryden, Michigan; Greg Schmitt, Noblesville, Indiana; Denis Schrank,

The 1939 Model H was the right tractor at the right time. It would share a wheelbase with the larger Model M so that Bert Benjamin's implements functioned interchangeably. This meant that farmers didn't have to purchase two sets of tools if they owned both an M and an H.

Batesville, Indiana; Neal and Shirley Stone, Wisconsin Dells, Wisconsin; Lawrence Terhune, Princeton, New Jersey; Martin, Marsha, and Matt Thieme, Noblesville, Indiana; Stew and Pat Thomet, Alto, Michigan; Bill Tyner, Westfield, Indiana; John Tysse, Crosby, North Dakota; Denis and Linda Van de Maele, Isleton, California; Mike and Paul Van Wormer, Frankenmuth, Michigan; John and Barbara Wagner, White Pigeon, Michigan; Gary and Judy Walton, Imlay City, Michigan; Norm and Ardeth Walton, Imlay City, Michigan; Warren and Janice Walton, Imlay City, Michigan; Louis, Linda, and Tim Wehrman, Reese, Michigan; and Bob, Kathy, and Randy Zarse, Reynolds, Indiana. Lastly, I am deeply grateful to my partner in life, Carolyn, for her love and encouragement and for so much more.

—*Randy Leffingwell*

ONE

1870–1919

PREPARING FOR THE FUTURE

The "Famous" engine turned a 12-inch (30.5 cm) diameter pulley. Operators moved a lever that shifted the engine to bring the pulley into contact with the 51-inch (130 cm) diameter friction drive wheel.

At the dawn of the twentieth century, dozens of inventors in the Midwestern United States were determined to put "Old Dobbin" out to pasture permanently.

Men with names like Olds, Buick, Maxwell, Ford, and even Studebaker—one of the largest horse-drawn wagon manufacturers—were hard at work to replace the horse with

This was John Steward's experimental tractor, at work in 1910. His clever rear-axle configuration allowed the operator and the tractor to ride level while running a lowered wheel in the freshly plowed furrow. The engine drove the front wheels by a chain. IHC records suggest they produced ten of these. *State Historical Society of Wisconsin, WHi (X3) 52038*

the automobile on the nation's roads. Likewise in Chicago, the rival Deering and McCormick companies merged together in an effort to dethrone the horse from America's farms. They founded the International Harvester Corporation (IHC) and their efforts eventually led to the birth of the much beloved Farmall brand some hundred years ago.

On August 12, 1902, McCormick, Deering, and three other harvesting equipment makers consolidated under the name International Harvester Company. The merger brought an end to the so-called "harvester wars" of the 1890s where McCormick and Deering engaged in intense competition that hurt every

maker. Deering and McCormick had attempted to merge their operations several times, including in 1891 and 1897, but mutual distrust had derailed the previous deals. The merger brought together complementary organizations.

Deering Harvester Company had a strong sales organization and owned steel mills and foundries. One of America's most advanced manufacturers of farming equipment, Deering had introduced the three-wheel, self-propelled "Automobile Mower" in 1894. Created by George H. Ellis and John Stewart, the prototype machine was powered by a 70-pound (32 kg), 6-horsepower two-cylinder engine.

McCormick Harvester was known for its production efficiency thanks to the adoption of precise manufacturing techniques and the concept of interchangeable parts. Instead of using skilled blacksmiths and machinists to make one or two products at a time, this "American system" used craftsmen to make patterns. Then semiskilled workers produced and finished thousands of parts while less skilled laborers assembled the final products dozens at a time. McCormick's annual farm implement production had risen from 17,500 units in 1880 to more than 100,000 by 1889.

An example of the intense competition and one-upmanship between Deering and McCormick involved the Paris Exposition World's Fair of 1900. Deering alone was nominated to represent American makers of harvesting machinery. Its centerpiece was an improved version of its Auto Mower prototype, which sported a 16-horsepower engine.

Although uninvited, Cyrus McCormick was determined to be among the 1,600 exhibitors at the Paris fair. His key engineer,

Edward A. Johnston, mounted one of his two-cylinder engines on a Bert Benjamin–strengthened cutter. Christened as "Auto-Mower," the similarly named McCormick machine outperformed Deering's Auto Mower during one test cutting a heavy growth of alfalfa.

Both Johnston's (McCormick) machine and Ellis/Steward's (Deering) version were innovative. It was an early form of a power take-off (PTO), which used one motor to propel the mower and power the cutters. While neither prototype was ever put into production, they showed the promise that machinery could soon replace the working horse.

After the Paris show, in February 1902, the chairman of United States Steel pitched a plan to McCormick, saying that his company should consolidate its operations with Deering to cut costs. McCormick accepted the idea. But only the outside intervention of George W. Perkins, a partner at J.P. Morgan and Company, paved the way for the Deering-McCormick merger.

The "House of Morgan" convinced the two harvester manufacturers that consolidation meant self-preservation. Perkins avoided the controversy of whether the Deering or McCormick families controlled the company by creating a voting trust where he was the tiebreaker. The newly minted International Harvester Corporation (IHC) named Cyrus McCormick Jr. as president and appointed Charles Deering, William's son, as chairman. Morgan and Perkins capitalized the new company at $120 million.

McCormick's growing sales force throughout Europe and beyond motivated Morgan to add the word "International" to the group's new name. IHC controlled nearly 90 percent of grain binder production and about 80 percent of the mowers in the United States. Sometime in 1905, IHC's Executive Council (EC)

created a product planning and review group informally called
the New Work Committee (NWC).

By 1906, the word *tractor*—which is attributed to Hart-Parr
Gasoline Engine Company—began replacing the term *traction
engine*. IHC fielded tractors under the Reliance name before
calling them Titans, after the mythical divine beings in Greek
mythology "of gigantic size and enormous strength."

The McCormick Works began making a large tractor named
the Mogul. The similar Titan and Mogul lines demonstrated that
the Deering-versus-McCormick rivalry died hard. One of the
most significant of these models was the prototype 1914 Mogul
20-40, the first mass-production application of the unitized tub.
It had an impact on later Farmall designs.

The Sales Department supported the bewildering assort-
ment of similarly named, identically performing machines.
Events in the next few years would end the confusion.

The creation of the new company came at a pivotal moment
in America, where tractors became very important. Farmers had
faced nearly three decades of hard times from the 1870s into the
1890s. They had a resurgence thanks to several factors, including the completion of the Transcontinental Railroad in 1869.
Additionally, America had more people to feed thanks to an
influx of a quarter million immigrants from Europe and Asia.

The 1911 22- to 45-horse-power Mogul. This was one of IHC's first products from its Chicago Tractor Works, the facility that began life as a big tent in the winter of 1910. By 1911 the reliability and popular demand of this tractor led IHC to manufacture 583 that year.

Fewer people, per capita, were working in agriculture by the time of IHC's foundation. In 1850, for example, some 70 percent of the 23.1 million people living in the United States toiled in agriculture. By 1910, the nation's populace grew to 92.2 million but only about a third, 33 percent, worked on farms. Working horses were unable to keep up with farming demands. To plow an acre (0.4 ha) of land with a horse took many hours.

International Harvester grew in Canada and America while expanding European sales. By 1910, to avoid European protective tariffs on agricultural implements and tractors, IHC opened factories in Sweden, France, Germany, and Russia. It was the first U.S. maker to sell a tractor in Russia. It also sold tractors in Argentina, South Africa, Austria, Mexico, Romania, Brazil, Turkey, Italy, Uruguay, Spain, Peru, Switzerland, Chile, Norway, and Serbia.

Just after Thanksgiving in 1910, the Executive Committee of IHC, which then included its general manager, Alexander Legge, looked at the market ahead that called for lighter tractors in the 8- to 12-horsepower range. Instead, Johnston had designed a 25-horsepower tractor that weighed 2,000 pounds (907 kg) less than the full-size Mogul. Using kerosene fuel and called the Mogul Jr., it was not the "light" tractor as envisioned, but it had a drawbar pull of about 4,500 pounds (2,041 kg).

Titan and Mogul 45s weighed 20,000 pounds (9,072 kg). The 18-horsepower, two-cylinder Universal tipped the scales at 9,000 pounds (4,082 kg) and featured sturdy 10-inch (25.4 cm) long pistons mounted to a beefy 3.5-inch (8.9 cm) diameter crankshaft. These behemoths ran steadily, but smaller farmers needed a tractor that would fit within their fences. Many Midwestern and Southern farms were only 5 acres (2 ha) and few were larger than 40 (16 ha). Most farmers wanted to work faster than a horse's walk, but they had already paid for their draft animals.

By 1910 farm journals urged tractor manufacturers to offer machines farmers needed: better built, more maneuverable, more reliable, easier to start, less cumbersome to operate, and less costly to purchase. It was not major manufacturers who answered magazine cries for lightweight tractors. Small makers such as Bull Tractor of Minneapolis, Minnesota, produced a 5,000-pound (2,268 kg), 5- to 12-horsepower tricycle. The Little

IHC's Ed Johnston conceived and directed manufacture of this two-cylinder giant. Early photographs show these tractors pulling as many as fifteen or eighteen plows, turning over a swath of earth 20 feet (6.1 m) wide or more. Such plow loads sometimes broke the frames of these tractors in the early days. The starting engine was one of Milwaukee Works' innovations with this big machine. One cylinder burned gas while the other compressed air. The operator pulled a lever and injected the air into the main engine to begin moving the large pistons.

Bull retailed for $335 and, by the end of 1914, the company had sold 3,800 of them; IHC slipped to second place in sales.

Robert Hendrickson and Clarence Eason of Wallis Cub took credit for introducing the "unit-frame" tractor that resisted twisting motions while in operation. This design improved torsional rigidity while cutting weight, complexity, and costs. Although only 660 Cubs were built over six years, Wallis led the way to other revolutionary products such as Fordson and Farmall.

While Harry A. Waterman at McCormick Works and Johnston made competing prototype machines in an apparent internal competition, Deering Works hired independent designer Harry C. Waite. He devised a finely engineered lightweight machine that immediately went into farm testing near Lewiston, Montana. He revised his three-speed transmission to two forward gears and reduced the tractor's weight, cutting initial $1,500 manufacturing costs to $850. His machine became the Steward-Waite tractor. The air intake for the carburetor was strained through water to protect from dust while all moving parts were covered.

During this time, Steward, who was IHC's expert on patents, saw reports of crawler-type tractors. While the idea of crawlers had been present in farming for more than twenty years, the name *Caterpillar* came from Ben Holt's machines. By the early 1920s, nearly two dozen manufacturers were producing crawlers or half-tracks.

IHC soon faced a new threat—an antitrust lawsuit by the U.S. Department of Justice (DOJ). By mid-1914, the DOJ was

hounding the conglomerate. IHC had assets worth nearly $173 million and ranked as America's fourth-largest company. As courtroom procedures occupied corporate time and resources, Henry Ford jumped into the tractor business.

In the fall of 1913, the three-year-old Ford Highland Park Plant had its first working assembly lines. Production time to build a car fell from twelve hours to one hour and thirty minutes. In 1914, Ford produced more cars than all its competitors combined—some 308,162 units. The son of a well-to-do farmer, Henry Ford also wanted a share of the tractor market. After learning of the Wallis Cub, Ford engineers Joe Galamb and Eugene Farkas created the Fordson tractor in 1916. It was built under a new company called Henry Ford & Son.

The number of companies claiming to be tractor makers grew from about 50 in 1913 to 165 by 1916. Among these companies was Samson Tractor of Wisconsin. General Motors purchased Samson in 1917 in an ill-conceived attempt to counter

Ford. Although sold under the GMC division, Samson failed by the early 1920s. While tractor makers produced heavy, ponderous units and others tried Ford-like automobile products, International Harvester followed a middle ground product strategy.

Overseas, World War I broke out in Europe in 1914. The need for food and cloth increased farmers' profits but also induced them to increase production. By 1916, the U.S. Department of Agriculture (USDA) reported nearly 34,371 tractors were working on American farms. Nearly 80 percent were two-plow machines rated to work at more than 2 miles (3.2 km) per hour. A tractor towing two, three, or four plows amazed newspaper writers who had never seen steam traction engines pull fifteen plows. Tractors would save America, the editors wrote, and they could save the world. North American makers turned out 62,742 machines in 1917, shipping 14,854 to Europe.

IHC's general-purpose Mogul was offered with grain drills, harvesters, and other implements, but it was still too clumsy and bulky to do delicate, precise row-crop cultivation. So in late 1915, engineer David Baker helped Johnston, Carl Mott, Philo Danly, and John Anthony complete the Motor Cultivator. IHC only tested two or three prototypes before producing the Motor Cultivator for 1917. Although the NWC ordered 300 to be built, IHC halted production at one hundred units for the year. By late 1917, the company recalled thirty-one cultivators so they could be retrofitted with a new engine governor, cooling fan, and heavier, cast-iron front wheels.

For the first time, IHC used its customers for final testing. Johnston had a good reason for hurrying production. He anticipated shortages in raw materials due to America's pending involvement in World War I. IHC reasoned that keeping existing machines in production was easier to justify than introducing a new one when rationing hit. Early buyers found the machine was top-heavy and prone to overturning, while its LeRoi engine was underpowered.

On April 6, 1917, the United States entered World War I. That same year, the DOJ's antitrust battle with IHC was nearly finished. The company had lost plants, sales branches, and products in Germany, France, and Russia, either to nationalization or bombing. To settle the lawsuit, IHC offered to sell three old-line harvesting machinery subsidiaries.

IHC's gear-drive 10-20 and 15-30 models represented a big advance in tractor engineering, manufacturing, and machine reliability. Here a prototype photographed on July 11, 1925, shows off an experimental articulated rear-wheel cleat for traction in soft soils and sand. *State Historical Society of Wisconsin, WHi (X3) 52028*

Sitting on IHC's new work committee or Executive Council in those days was akin to living in nearly constant turmoil. The corporation's new channel-frame tractor, the four-cylinder Mogul 8-16, conceived in 1914 was reborn later in 1917. Its problems required two engine transplants. The International 8-16 (the name ultimately given it on October 12, 1917) offered farmers America's first production power take-off. The 8-16 was used to test four-wheel and six-wheel drive and crawler tracks. Later—in June 1919—the tractor was given extra-wide, reinforced steel wheels for rice fieldwork.

left: The 1915 Mogul 8-16. IHC introduced this model in 1914. Its arched front framework allowed it to turn extremely tightly, for its time, within a 20-foot (6.1 m) radius.

right: The 1917 Mogul 10-20. As successful as the 8-16 was (IHC sold about 5,000 of them in 1915 and more than 8,000 in 1916), Ed Johnston felt they needed more. He increased engine-operating speed and added a second forward speed. IHC offered optional plow guides for 8-16s and continued with the new 10-20s. The horizontal single cylinder ran a planetary gear transmission to power the rear wheels using a left-side chain. The transmission provided only one forward and one reverse speed.

In September 1917, IHC engineer Bert Benjamin was in Nappanee, Indiana, watching three of his International 4-horsepower "Binder Engines" harvest hemp, a product in great demand by the military for a variety of uses. A Titan 10-20 pulled one binder, four horses pulled another, and an 8-horsepower competitor pulled a third through very tall hemp. Hemp dust and leaves choked the towed binder's auxiliary engine, which lost a quarter of its power. Yet, the tractors suffered no ill effects from the dust. Benjamin concluded that the tractor engine should drive the binder by PTO shaft. It took the company about a year to field a prototype PTO that operated the cutter bar of a mower attachment and a sweep rake lift.

In 1918 wartime uses of steel and other material threatened all domestic industry. The U.S. Government Priorities Board limited total production by all manufacturers to 315,000 tractors.

In fact, the final count reached only 132,697, and the armistice in November brought an end to allocations in December. More than one hundred new companies entered the tractor business. Of these more than 250 tractor makers, only 98 produced a single tractor.

One of the tractor competitors was the Fordson. Henry Ford considered his real competition to be draft horses, so he asked IHC and Deere & Company for advice. Legge was then part of the U.S. War Production Board. When approached by Ford, Legge loaned the expertise of Benjamin for several months. While Benjamin designed a Fordson line of implements, he also studied Ford's assembly-line methods. When Ford dropped plans to sell implements, IHC provided them to Ford dealers instead.

Ford supplied 4,260 Fordsons to England before producing them for American farmers starting in April 1918. All the publicity created a demand for the small Fords, but the company's American production got off to a slow start. Other manufacturers began to fill farmers' needs.

As the wartime economy tightened, IHC Manufacturing's Henry Utley endorsed a two-tractor, large- and small-machine plan while the Motor Cultivator was the distant third in priority. The 1918 machines could run at higher engine speeds without overheating. Tractor Works assembled 160 units by July, ending the year at 301, including rebuilding 1917 models. This slow pace resulted from steel shortages during World War I; the government allowed manufacturers 75 percent of their 1917 quantities.

In August 1918, about two months before the armistice, the NWC terminated the Motor Cultivator. The company delivered 213 units in 1918, a number that included 67 leftover 1917s redone as 1918s, and 84 units in 1919. The final report judged the Motor Cultivator "could not be produced at a cost which it was estimated the farmer would pay." IHC had wanted farmers to buy a "general-purpose" tractor (large or small) for plowing and harvesting and a second machine for cultivation.

Prior to peace in Europe on November 11, 1918, IHC surrendered to the U.S. Department of Justice. The painful terms of its agreement included releasing 4,778 dealers, many of whom joined John Deere. With the lawsuit out of the way, IHC began growing its tractor business.

above: The 1918 International 8-16. This was a great idea that led to other great ideas, including power take-off and chain final-drive systems. In addition, the radiator and fan were located behind the engine and below the fuel tank. Engine heat helped vaporize the fuel. It probably helped vaporize operators' feet as well. Ultimately, its channel frame was not strong enough, and this tractor never was a great success.

opposite: IHC's four-cylinder engine ran at 1,000 rpm using a magneto for ignition spark. Early engines suffered from inadequate lubrication.

INTERNATIONAL
8-16

1918 INTERNATIONAL
MOTOR CULTIVATOR

The company had its own assembly line running in 1918 to build the International 8-16. The winter of 1918–1919 was mild, enabling farmers to work later in the fields in the fall and go in earlier the next spring. Europe still needed food and cloth, and farmers expanded throughout 1919 and 1920.

IHC ended the Mogul 10-20 in 1919, yet it continued the Titan 10-20 until 1922 for a number of reasons. Ford's competition and frustrations with the smaller 8-16 tractor left the company little choice.

In June 1919, Johnston took Motor Cultivators to Blue Mound, Illinois, for the first strictly cultivator show in the Corn Belt. Avery, Moline, and Allis-Chalmers demonstrated machines. J. I. Case Plow Works introduced a rail-frame 12-horsepower cultivator. Far lighter than a Wallis Cub, it signaled a new direction in machinery design.

After the Illinois show, Benjamin envisioned a highly adaptable "Combined Tractor Truck" of two- to four-ton capacity, powered by a 15- to 25-horsepower kerosene engine. It would carry "a combined harvester-thresher, power direct from the engine,

INTERNATIONAL
HARVESTER CORP.
TRACTOR WORKS
CHICAGO
ILL.
INTERNATIONAL
INTERNATIONAL
8-16
KEROSENE
TRACTOR

with speed independent of the tractor, [and] could carry and operate a grain binder and shocker, a corn picking device with a box for loads, a hay loader and rack for hauling hay, or a water tank, pump, and sprinkler device for fire protection on the farm."

Although IHC had stopped manufacturing the Motor Cultivator, Mott and Johnston continued development efforts to create a "cultivating tractor." In late 1919, Johnston asked for name ideas. Ed Kimbark suggested "Farm-All," as documented in the November 10, 1919, Tractor Works records. By early 1920, they dropped the hyphen and the name became Farmall.

left: The 1920 Titan 10-20. It lacked the sophisticated appearance of International's 8-16, but the Titan became the standard bearer for IHC when Henry Ford and his Fordson declared a price war. With new machines in the pipeline, IHC's Alex Legge could afford to discount Titans and match Ford's price, cut for cut.

The Titan's two 6.5 x 8-inch (16.5 x 20.3 cm) cylinders gave the tractor enough power for three plows. Its two-speed gearbox provided a choice of 2.25 or 2.875 mph (3.6 or 4.6 kph) forward, and a single 2.875 mph (4.6 kph) reverse.

right: The 1920 Titan 10-20 was another successful machine for IHC. In 1920 alone, the company manufactured 21,503, and output from 1915 through 1922 totaled 78,363 units. This was a three-plow-rated machine compared to the two-plow-rated Fordson.

opposite: The 1919 International 8-16. The tapered nose appealed to orchard owners who saw its ability to slip under branches. Its radical looks, however, put off some conservative buyers. At the same time, it reminded others of contemporary automobiles.

The 4-inch (10.2 cm) bore and 5-inch (12.7 cm) stroke engine went through several incarnations as IHC struggled to get its problems resolved. The tractor ended up 600 pounds (272.2 kg) heavier than the similarly powered Fordson, and it cost more, further harming sales.

TWO

1920–1929

FROM DOUBTS TO SUCCESS

The 1920 Prototype Corn Harvester. Tractor Works engineer David Baker worked with three colleagues to create the Motor Cultivator in late 1915. While it never succeeded completely, it became the test bed from which important ideas grew. By December 1920, the machine had shed its cultivators and adopted a double-row corn harvester from Bert Benjamin's McCormick Works. *State Historical Society of Wisconsin, WHi (X3) 52021*

John Steward and Harry Waite together had planted a seed inside International Harvester Corporation to create a machine capable of doing all the jobs routinely performed by horses. Though Steward had died of heart failure in 1915, other IHC engineers, including Edward Johnston, David Baker, Carl Mott, Philo Danly, and John Anthony, continued this quest of innovation, which led to the development of the first Farmall tractors.

Tractor Works created a new tractor from Bert Benjamin's research of farming methods and implement design. One tractor by Baker used a Waukesha engine fitted with a reversible operator's seat and a transmission with three speeds, both forward and reverse. These new ideas got little support. The post–World War I recession, slow sales, and Motor Cultivator losses left little money to hand-assemble more than two Farmall prototypes and test them. Prototypes often cost fifty times what a finished production version can cost to manufacture. The Model A Farmall, using an L-head truck engine from Akron, appeared around February 7, 1920; the Model B arrived on June 30, using a new engine that carried through to production.

Benjamin campaigned for the reversible Farmall. He wrote general manager Alexander Legge on October 15, 1920, that the prototypes performed eleven separate farm operations using a single operator. Adopting automotive-type engines, he priced the machine for $900. The $1,000 International 8-16 did only four tasks with one individual who had to rely on horse teams to do the other seven jobs.

While backing the Farmall, Johnston noted that a modest experimental program would cost $150,000 to $300,000 to build five tractors and implement sets by hand. Still, he advocated enlarging the test fleet. An early detractor was J. F. Jones, the Chicago office sales manager, who said that farmers wouldn't take to it because it was "built on exactly the wrong lines." He suggested replacing it with something heavier.

Legge remained neutral. He and fellow hobby farmers Harold and Cyrus McCormick Jr. recognized the Farmall's value, but IHC remained at war with Ford, and cash was tight after the Motor Cultivator disappointment.

During 1920, the USDA reported that U.S. manufacturers sold 162,988 tractors. Fordsons constituted 35 percent of the

market, while 15 percent came from IHC. Tractors worked on 229,334 farms that year, representing just 3.6 percent of all operations, meaning that there were many potential customers. Then crop prices dipped in 1921 as European farmers got back to work. Tractor sales dropped to barely 35 percent of the peak reached during 1920.

Around Christmas 1920, one of Benjamin's engineers, C. A. Hagadone at McCormick Works, sketched a lighter-weight version of the Farmall at about half the 4,000-pound (1,814 kg) approved model. On January 21, 1921, the NWC canceled the heavy prototypes and ordered "two of the modified, lightened Farmalls." This was the first time Ed Kimbark's suggested name appeared in official IHC papers.

The lightweight Farmall ran in one direction, steering in front while its two powered wheels pushed from the rear. Tractor Works enclosed final-drive housings and moved the cultivators to straddle the single front wheel, which allowed farmers to cultivate without damaging crops within the contours of a row. Still, McKinstry's Sales Department found nothing in this new, fast tractor that it could sell as an advantage over horses. Researching Farmall's potential benefits to farmers, Utley looked at the costs of a 160-acre (65 ha) hog farm. Replacing horses with a Farmall turned feed lands into cash crops—a potential net income of $3,500—versus $3,000 for farmers who had a Fordson but still needed six or eight horses for other activities.

Of the 186 companies claiming to be tractor manufacturers in 1921, IHC considered Ford its only challenger. Legge held meetings to consider Fordson and Farmall developments. Although a depression in the farming sector depressed tractor sales, Legge delivered his verdict at the July 21, 1921, session. Farmall implement development would continue. Tractor Works would expand development to a hundred Farmall prototypes for 1922. Cyrus and Harold McCormick championed Farmall in the shadows. In late December, Legge ordered manufacture of twenty

Farmalls with mowers, corn planters, and several other tools.

The market war intensified in January 1922 when Ford chopped the Fordson's price by $165, from $790 to $625. IHC followed by dropping the $1,200 Titan 10-20 to $900; the $1,150 International 8-16 went to $900; and the 15-30 decreased from $2,300 to $1,750. By year end, Ford had sold half of the 34,000 tractors produced nationally and IHC had sold one-quarter.

Horses were cheap and crop prices remained low. The market stagnated until February 5, 1922, when Ford cut the Fordson price by $230, from $625 to $395. IHC followed, reducing the 8-16 another $230 and the Titan down to $700. Other manufacturers followed the price cuts.

On February 10, 1922, Legge reconvened the Special Tractor Conference. The battle with Ford was at full pitch. Benjamin proposed reducing Farmall power and weight. Chicago sales manager and naysayer Jones still saw no value in any Farmall—heavy, light, or cheap. He urged Tractor Works to find ways to "bring the cost of the new 10-20 tractor down."

Benjamin, Danly, and Baker improved their design. Then a late and very wet spring delayed the planting season. Many farmers found that only a tractor, not horses, were able to work the fields in time to make a good harvest. Tractor sales had a rebound.

IHC's directors unanimously elected Legge president on June 2. Harold McCormick became chairman of the board. McKinstry became president of IHC of America, the sales arm of IHC. Irked by poor prototype quality, Legge wrote Utley on June 23, 1922, chastising Manufacturing for undermining Farmall development. "About the only thing we have demonstrated so far is that we have done a very poor job in putting [Farmall prototypes] together, which suggests that you should strengthen your engineering staff to the extent that you avoid letting out into the field things that have to be rebuilt."

On July 24, Legge and Utley went to Hinsdale to watch Benjamin's light, cheap Farmall in action. What they saw converted them both. Benjamin's new machine, lighter by 800 pounds (363 kg), provided three forward speeds but only one reverse. Other changes included improved steering and placed the cultivators at the front of the tractor. The NWC killed the large Farmall. After a demonstration of "Benjamin's tractor" in August, many detractors, including Jones, decided to proceed with development.

By late February 1923, Legge and the EC unanimously approved assembling twenty-five Farmalls by hand at the Engineering Department. Johnston and Legge calculated mass-production costs and found it was nearly the same as the 10-20 Gear Drive. The company got its Farmall trademark registration in Washington on July 17, 1923.

Twenty-six prototype models were completed by August 9 and were sent out for tests. Four tractors remained at Hinsdale, one went to Cyrus McCormick Jr.'s farm in Wheaton, Illinois, and another to Utley's in Downers Grove, Illinois. Johnston shipped thirteen units to branch houses in Georgia, Mississippi, Tennessee, Texas, and Wisconsin for evaluation by farmers. Almost without exception, the results were good and the tractors held together. Reports to Legge said it had "plenty of power" and "splendid ground-gripping qualities."

Benjamin interviewed operators on the Durham Farm in Wayne, Illinois, where one hand told him, "In cultivating young

The 1923 Farmall Prototype. Neal Stone's rare prototype glistens in the afternoon sunlight. By the time IHC's engineers had reached this stage with tractor development, they pretty much had everything right.

Tractor Works engineering documents referred to this as, first, the Combined Tractor-Truck, and later the Cultivating Tractor. Engineering secretary Ed Kimbark first wrote the name Farm-All on November 10, 1919.

corn 4 or 5 inches [10 or 12.5 cm] high, we can do a better job with the Farmall because it handles so easy . . . With a two-row cultivator pulled by three horses, we have to pay so much attention to the horses that we cannot do a good job cultivating."

In October, McKinstry recommended producing a hundred Farmalls for 1924, while Legge and Johnston advocated for more due to the positive reaction from southern testers. The only drawback for the machines was their unattractive appearance. But, one cotton farmer in Texas noted: "It's homely as the devil, but if you don't want to buy one you better stay off the seat."

In 1923 Ford reached its peak, producing 101,898 tractors and capturing 76 percent of the market. It left 9 percent, or 12,057 machines, to IHC. All the other seventy-three producers combined made up the remaining 15 percent. In 1924 Ford sales slipped to 70.9 percent, or 83,010 tractors, while IHC grew to 16 percent, with 18,758 machines. Dozens of competitors ceased production during 1924. Those who remained only garnered 14 percent of the total.

IHC's tractor improvements for 1924 included enlarging rear-axle diameter, increasing drawbar fastening size, fabricating steel starting gears, making main frames of rolled tubing, and strengthening bull-gear housings. Tractor Works enlarged kerosene fuel tanks to 13 gallons (49 L). Selling for $825, the early Farmalls lost substantial sums for IHC because they were hand

McCORMICK-DEERING
FARMALL

FARMALL

built. The EC accepted this as promotional expenses. In late February 1924, the NWC and EC ordered two hundred Farmalls built to expand testing. One of the early models (QC-501) was sent to farmer Roy Murphy in Taft, Texas, who also received the first "skeleton" wheels; these open metal frameworks soon became known as Texas wheels.

Benjamin spent two weeks in Taft, where Murphy asked him for a four-row cultivator. Working with a local blacksmith, Benjamin fabricated a prototype cultivator. Excited with the results, Murphy called in his neighbors to watch as he covered 100 acres (40.5 ha) in fourteen and a half hours, compared to 18 acres (7.3 ha) the previous ten-hour day with his original two-row cultivator. Benjamin took plans to McCormick Works for production for South Texas cotton farming. The NWC approved production of fifty units for further testing.

By September, Tractor Works had turned out 205 of Benjamin's tractors with modifications that included a

simplified steering gear and gear shift, a muffler for the exhaust pipe, and a strengthened transmission case. When Jones, the reluctant convert now proposing manufacture of three hundred units, wondered whether the Farmall might harm 10-20 Gear Drive sales, Johnston answered, "There has been a constant cry for a tractor to meet Ford's. It is impossible to meet Ford on price; therefore, we have to produce something of greater utility to justify our price. The Farmall will do this. If the Farmall is the tractor that will kill the 10-20, it would be far better if we ourselves kill it."

With the last critic at peace, the NWC and EC agreed to manufacture 250 units. Dealers in some branches were told to plug 10-20s and 15-30s, delivering Farmalls only when buyers insisted. Sales told other branches that they could not get one at all. Farmall sales were pushed in the cotton South to pick up new customers. Studies showed that the new tractor would cut cotton production costs from $110 a bale with mules to $83.

Sparring with Jones at one meeting, Benjamin said: "You see that operating by Farmall . . . makes it possible to save the (cotton) crop for the United States instead of losing it to Egypt, southern Russia, India, the Argentine, . . . producing [there] with cheap labor at $95 a bale." In April, Jones went to Texas to see the tractors in operation for himself.

Outside of San Angelo, Jones watched a demonstration with P. Y. Timmons, manager of tractor sales; Jim Ryan, Houston branch manager; Joe Foley, Dallas's manager; Guy Fisk, Amarillo's branch manager; several implement engineers; and Benjamin. As the afternoon wound down, the group discussed the Farmall. Timmons reflected Jones's lingering concern that it might adversely affect 10-20 sales. Fisk, Foley, and Ryan agreed they "had little use for [10-20s] in cotton country. They were not adaptable to all row-crop operations."

Ryan didn't hesitate: "If you don't adopt it for production, we'll organize a company in Houston and build it down here." Jones, parroting Johnston's earlier comments, said: "If anyone was going to build a tractor that would affect the sale of the 10-20, let's do it ourselves."

Production plans grew from having 1,000 Farmalls ready by October 1925 to 2,500 units. McKinstry added another 1,000 for the 1926 season. Good crop harvests created more interest in the tractor. In September 1925, the EC ordered Tractor Works to

FARMALL

solve the two chronic Farmall complaints about broken studs on the differential and troubles with the steering.

In the sales war against Ford, IHC dealers seized every opportunity to show the Fordson's limitations. Ford's market share eroded to 64 percent in 1925 while IHC's rose to one-fifth of the 164,097 tractors produced nationwide. Ford chose not to improve the tractor. Yet the Fordson revolutionized farming, proving that a small, lightweight tractor could be mass-produced, sold cheaply, and replace the horse.

On March 19, 1926, Utley reported that Tractor Works was making eight Farmalls a day, having met outstanding orders for 1,708. IHC priced them $100 higher than the 10-20s to emphasize their greater potential. The order climbed to 2,954, and Utley expected daily production to reach fifteen by July 1.

IHC was able to increase production because it opened the Farmall Works in June 1925. The company had previously acquired this plant in Rock Island, Illinois, from Moline Plow Works. By November 5, 1926, near the end of another record crop harvest, Rock Island production settled at twenty to twenty-five units per day. Suddenly the Midwest was hit with hard rains and high winds with storms lasting for days. Harvesting and threshing stopped and fields flooded. Some states' grain harvests were entirely ruined. Crop prices slipped rather than rising from destroyed supplies. Predictably, tractor sales fell. While U.S. makers produced 178,074 tractors in 1926, they sold only 122,940 (46,441 as exports); 4,430 U.S. sales were Farmalls. McKinstry hoped 1927 would reach 6,600, while district sales managers believed 7,500 was more realistic. Legge took a leap

left: IHC's tractor engineers made missteps as they invented the next generation of IHC's tractors. It took implement engineer Bert Benjamin to crystallize in the engineers' minds what he believed farmers wanted.

right: The 200,000th 10-20 Gear Drive. On June 4, 1930, production stopped for a few moments for a celebration. Despite dealer fears that Farmall sales would sour the market for standard-front 10-20 and 15-30 tractors, IHC manufactured this landmark wide-front machine sixteen years after Farmall production began. *State Historical Society of Wisconsin*

of faith and he guessed right: Rock Island manufactured 9,502 throughout 1927.

In 1927 the U.S. Supreme Court refused to reopen the DOJ case against International Harvester, ending one of the two great challenges to the company's existence. The Fordson tractor market share slipped to 31 percent during 1927 and 1928 while IHC's rose to an equal amount in 1927 before gaining 62 percent of the business in 1928.

After assembling 8,000 Fordsons by mid-1928, Ford ceased U.S. tractor production and transferred all Fordson tooling and manufacture to Cork, Ireland. Meanwhile, IHC introduced its TracTracTors, which were crawlers based on the McCormick-Deering 15-30 and 10-20. But it was the

Farmall that would revolutionize farming as thoroughly as the Fordson had.

Tractor Works introduced a narrow-tread model for export on May 5, 1927. Production for 1928 reached 24,899 Farmalls, and 35,517 in 1929. But the "Roaring Twenties"—a time of economic boom—was coming to an end due to numerous factors, including the bursting of the Florida real estate bubble and speculators driving up stock prices.

In 1929, President Herbert Hoover created the Federal Farm Bureau to help stabilize farm market prices. He asked Alex Legge to join the bureau's new board. Although resigning from IHC's presidency, Legge remained on the company's board of directors. He occasionally returned to Chicago for company meetings

a legend is born

tRACTORS PRIOR TO 1924 were tailored for plowing and belt work. The rest of the farmer's daily tasks were not yet addressed by mechanization. Implement makers had looked into the possibility of engine-driven mowers, rakes, and cultivators, but nothing came of their experiments. It took the onslaught of the Fordson to inspire the engineers at International Harvester to dust off their designs and get serious about an all-purpose tractor.

The Experimental Department at IHC was not caught without their homework done. In fact, they had gone so far as to file for a patent on a machine they called a motor cultivator. Then realizing that farmers could not afford a single-purpose engine-driven machine, they worked to combine the conventional tractor with the motor cultivator. The result was a spindly looking machine with wide-spread large rear wheels with two smaller closely spaced front wheels under the engine. It was high enough and had good operator visibility for cultivating crops. The machine was equipped for draft or belt work, plus it had a rear PTO output for powering harvesters. The result was unofficially called the "Farm-All," which later morphed into Farmall. The Farmall could replace all the horses on the farm.

The first Farmalls were sold in 1924, although they could have been called field test models because they were closely watched by IHC field men. Production in earnest started in 1925.

In 1932, the Farmall became a series of two models, the modernized F-20 and the larger F-30. To differentiate between the first Farmalls and the F-20 versions, the first or basic Farmalls picked up the nickname "Farmall Regular." There was also a commercial golf course model called the Fairway. The Farmall went on to become one of agriculture's most significant tractors. — *Robert N. Pripps*

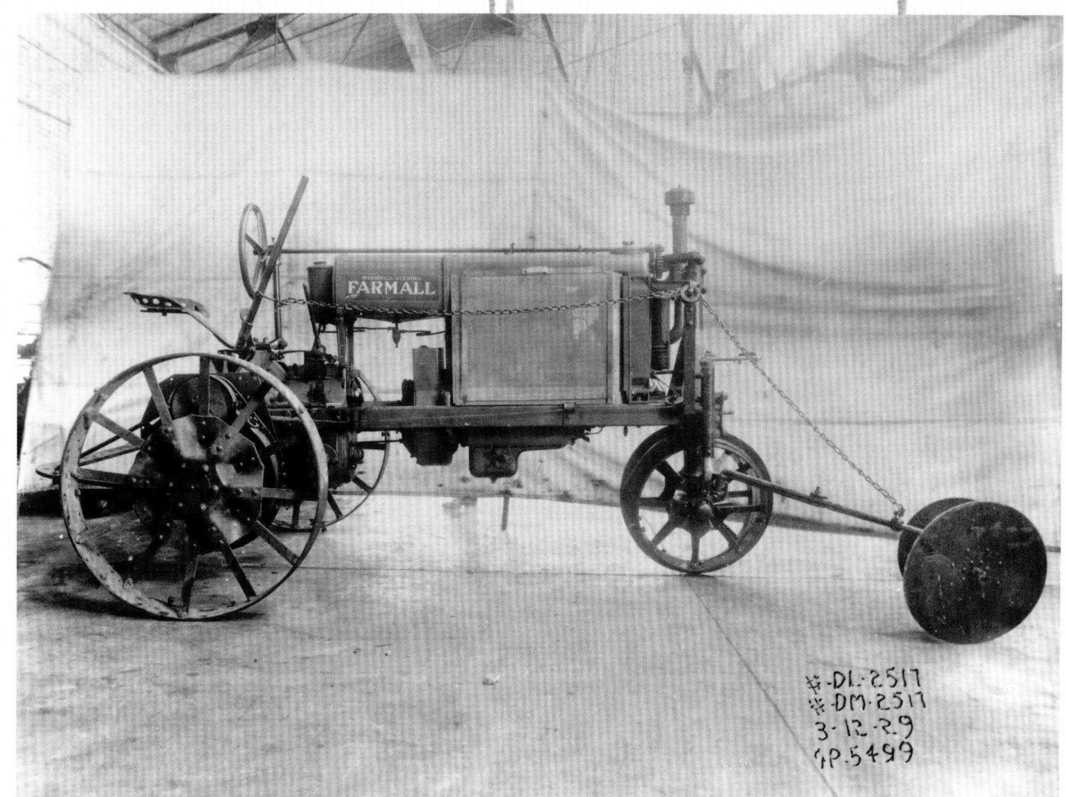

The 1928 Farmall prototype with Plow Guide and Engine Cover. This Farmall shows off a few options, some destined for production, others not. IHC had marketed plow guides for its tractors for decades, so the "Steering Device DL-2517" is no surprise. The single front wheel and the engine cover are unusual. This photo, dated March 12, 1929, shows one of IHC's 1928 test "mules," #23397. *State Historical Society of Wisconsin, WHi (X3) 51990*

left: The 1930 Farmall NT. Tractor Works introduced a narrow-tread model for export on May 5, 1927, mainly to Argentina. IHC had acquired Moline Plow Company's Rock Island, Illinois, tractor factory in 1924. This allowed IHC to accelerate production and begin special models when production started there in June 1926.

Narrow-tread (NT) models received an offset rear hub. This collapsed rear tread width from the standard 74 inches (188 cm) down to about 63 (160 cm). Documents in archives suggest an even narrower version provided 57 inches (144.8 cm), but this appears more commonly on Fairway tractor specifications.

right: This original, unrestored NT shows little use and extensive care during its lifetime. IHC began using its E4A magnetos on Farmalls after 1926.

opposite: Production in 1930 reached the peak, at 42,093 Farmalls completed at the Rock Island plant. On April 12, IHC celebrated completion of the 100,000th Farmall.

that particularly interested him, especially when his efforts in Washington made little progress.

Starting in September 1929 and continuing through November, the New York stock market would lose half of its value, plummeting from $64 billion to $30 billion. The stock market crash marked the beginning of the Great Depression and new challenges for International Harvester and Farmall.

McCORMICK-DEERING
FARMALL

THREE

1930–1935

BECOMING A BIG FAMILY

The 1936 F-20. The F-20 was the successor to IHC's groundbreaking row-crop Farmall. With internal engine changes and new exhaust manifolds, engineers initially pushed 20 percent more power out of the engine. Test results using various other fuels often showed more of an increase than that.

International Harvester Corporation staked its claim as North America's dominant tractor maker in the 1930s as it set a number of milestones despite the Great Depression. Daily Farmall production totals reached 200 units by January 27, 1930. By year's end, IHC had made 42,093 tractors while the U.S. Census Bureau counted 920,378 tractors on farms.

McCORMICK-DEERING
FARMALL
INTERNATIONAL HARVESTER COMPANY

On April 12, Rock Island's Farmall Works manufactured the 100,000th copy of Benjamin's tractor. Within two months, the corporation celebrated again as the 200,000th 10-20 Gear Drive, with Cyrus McCormick Jr. at the wheel, rolled out of Tractor Works.

As the January 9, 1930, issue of *Farm Implement News* reported, "Three-plow standard type tractors have added . . . to their stature . . . because the Great Plains wheat farmers discovered that they could handle another 100 acres [40.5 ha] or so with wheatland disk plows, duck-foot cultivators, rotary rod weeders, and combines, if their three-plow tractors only had enough . . . power to pull an extra foot or two of disk or to carry the combine up slopes."

In early 1930, NWC notes and EC minutes mentioned the Gas Power Engineering Department (GPED)'s "Increased Power Program." But they weren't describing changes to ten-year-old 15-30s and 10-20s.

"Progress," Johnston told EC, "has put our competitors in a position to increase the horsepower for the size of engine and to improve the fuel consumption. We are suffering in the trade." He urged the production of a more powerful Farmall, and even a smaller one. EC members approved three sizes, counting the current Farmall and the proposed "increased-power Farmall" as one. The second was an intermediate Farmall, using an increased-power engine. Third was a large Farmall designed to use the increased-horsepower 15-30 tractor engine.

The power increases came from a new head, intake manifold, and piston design without changing bore or stroke. To Johnston,

the term *increased power* meant an "improved tractor." Upgrades included a water pump with better thermostatic control to improve cooling, strengthened frames for Industrial Model 20s, and a wide tread for the Farmall.

On December 1, 1930, Legge redefined the experimental Farmalls. The Increased Power model would handle two plows. The Intermediate, based on the improved 10-20, would run three plows. The Large 15-30-derived Farmall would pull four. Baker suggested fitting four-speed transmissions into the Large and Intermediate models and simply dropping the "Regular" model. A newcomer to NWC meetings, John L. "Mac" McCaffrey, IHC's

Engine output increased because Ed Johnston's engineers designed and tested new cylinder heads, intake manifolds, and pistons. He added a water pump, which greatly improved engine cooling and lubricant life.

left: This F-20 was the middle prong of Ed Johnston's three-way attack on IHC's competitors. Using an Increased Power engine from the 10-20, this became the new Intermediate Farmall. *State Historical Society of Wisconsin, WHiM90-048 F20*

opposite: The 1930 Improved Power Farmall Prototype. Ken Holmstrom's prototype showed subtle features that distinguished it from late-production Regulars. The canted front wheels were the first giveaway. Internally, the improvement in power came from new engine pieces.

assistant manager of domestic sales, disagreed. The Naming Committee designated the two-plow model as the F-20, the Intermediate was the F-30, and the Large Farmall was the F-40.

By 1931, the Depression, made worse by a long drought in 1930 that destroyed crops, caught up IHC. Farmall sales collapsed by two-thirds to 14,093 units for the year. Competition fell away too. Where there had been 186 registered tractor makers in 1921, by 1930 only 33 remained. Yet, Benjamin and Johnston recognized the need for both smaller and larger farm machines.

Frustrated by his inability to help farmers and turn around the world economy, Legge left the Federal Farm Board in early March 1931. He returned as president of IHC and immediately got to work. At an IHC special conference in Phoenix, Arizona, Legge noted that the company had to make machines for the average farmer. "We must not make our tractors too heavy, too high in cost, and too expensive in operation," he said.

Benjamin proposed returning to the Two Tractor Plan. He said that a 24-horsepower Increased Power Farmall, along with the Intermediate Farmall, would take care of 90 percent of IHC's current business. The other 10 percent of sales came from California, where IHC's new crawlers would fill farmers' needs. The EC approved large-scale production of the F-30 Intermediate Farmall and the Increased Power F-20.

On July 14, board chairman Cyrus McCormick Jr. signed off on improvements to the "Regular" Farmall, including raising output by 3 horsepower, adding a four-speed transmission,

McCORMICK-DEERING
FARMALL
INTERNATIONAL HARVESTER COMPANY

McCORMICK-DEERING
FAIRWAY
INTERNATIONAL HARVESTER COMPANY
GASOLINE

and enclosing the steering gear. Later, Johnston asked Baker to design a new one-plow Farmall, an F-10, with their modified unit-frame at the rear for the engine, transmission, and running gear while extension rails supported the front axle.

In January 1932, the NWC dealt with variations on old themes: modifying the regular narrow Farmall as a new Fairway tractor by replacing the front wheels that often cut into the bunkers with a wide front axle. Johnston created a similar configuration as a wide-tread front axle for the F-30 narrow-tread tractor as a no-additional-charge option. He used wheels off the 10-20 and the wide-front versions of the Farmall F-30. By April, his engineers had devised a wide front axle for the F-20 that would fit the Regular Farmall.

Baker continued work on the F-10 one-plow tractor. Tractor Works released the first of these smaller semi-unit-frame models as the F-12. Competition nipped at the Farmall's heels. The Regular's production totals dipped to 3,080 for 1932, though IHC tractor prospects were bolstered by production of 2,500 F-20s as well as 1,500 F-12s.

The trend toward inflatable rubber tires caught up with IHC. Through 1932, GPED's Sperry urged the NWC to keep up with competition, which was offering low-pressure tires with inner tubes rather than zero-pressure, solid rubber tires.

"It is entirely possible that pneumatic tires may be developed to meet many agricultural operations as they are now

left: IHC relied on its E4A magneto from model introduction in 1942 for another few years before replacing it with the F4.

right: Tested at the University of Nebraska in May 1933, the F-12 developed 10.1 horsepower at the drawbar and 14.6 off the pulley or PTO. It weighed 2,700 pounds (1,225 kg), compared with 3,950 pounds (1,792 kg) for the F-20 and 5,300 pounds (2,404 kg) for the F-30.

meeting industrial tractor needs," Sperry observed. "Allis-Chalmers are advertising pneumatic tires on farm tractors. There is a possibility that these tires may cut into crawler tractors sales: Caterpillar is experimenting with pneumatic tires on wheel tractors."

The NWC approved low-pressure pneumatics for the increased-power 10-20s, W-30s, and the Farmall Regular as well as the F-12 tractors. They offered F-12s as Orchard, Industrial, and Fairway versions. Sperry recommended providing a fourth, much higher speed gear, determining that 10 miles (16 km) per hour now seemed sensible.

In early 1933, the NWC addressed ongoing problems large and small: magnetos and impulse starter couplings for four- and six-cylinder Farmalls; a clutch-release hitch that disengaged if the plow hit something; low-pressure pneumatic tires for the F-12; worm-steering gears for the F-20 and F-30; kerosene engines for the F-12; continued pressure on GPED for diesel engines; electric starters; relocated air cleaners; new engine crankcases to provide better lubrication to the bottom end and tops of the F-20 engines; corresponding widening of the tractor frame with hood and fuel tank to match; and elimination of engine side doors to improve the appearance.

little brother farmalls

tHE DIMINUTIVE FARMALL F-12 made its debut in late 1932. About 550 were equipped with a Waukesha engine. This was replaced by an overhead valve engine of Harvester's own design. The F-12 was a scale model of the F-20, but rather than the drop box final drives, the F-12 used straight-splined axles so that infinite rear wheel spacing could be accomplished. At first, a single front wheel was used, but soon dual tricycle and wide fronts were made available. Also available were a central rear PTO, rubber tires, and a mechanical implement lift. Gasoline fuel was standard, but a kerosene manifold and starting gas tank were an extra-cost option. A Fairway model for golf courses and airports was added.

The standard-tread version of the F-12 was the W-12. It was the same mechanically but built lower and with fixed tread widths. Of the W-12 configuration, additional variations emerged: O-12 (Orchard) and I-12 (Industrial).

Production of the F-12 Series ended in 1938 with more than 123,000 delivered. It was replaced by the almost identical Model F-14. The only visible difference from the F-12 was that the steering wheel and shaft were raised to a more comfortable angle and position. Increasing the engine speed from 1,400 to 1,650 rpm gave about 14 percent more horsepower, enough to give the F-14 a two-plow rating. It was available in either gasoline or kerosene versions, and on steel or rubber. Wheel weights were an option when rubber tires were specified. A hydraulic implement lift was an option. W-14, I-14, O-14, and Fairway 14 versions were similar to those of the F-12 Series. Interestingly, the 12 and 14 Series tractors were the only ones made by IHC at the time to use fuel pumps. — *Robert N. Pripps*

"The committee," Kimbark read in a letter from Harold McCormick during a June 20 NWC meeting, "are impressed with the advantages of the construction used in the F-12 Farmall, compared to the Regular, the F-20 and F-30. Designing this form of chassis, having high wheels and one-chamber gear case, should be combined with designing modern, higher-speed, four-cylinder engines for tractors of the two-plow and three-plow sizes."

Sperry went pale. Was this an order to redo the entire line? His portion of GPED was overextended as it was. The F-12 meant a lot to him, but he had no money, no personnel, and no time to develop new tractors, even ones based on his own idea. The high-wheel design presented obstacles to attaching existing Farmall implements. The changes required to existing implements outweighed the advantages. Johnston voted against another hurried project.

The four-plow tractor idea returned in discussions on September 11, 1933. Based on semi-unitized F-12s with speeds ranging from 2 to 20 miles (3.2 to 32 km) per hour, the NWC now wanted this with the diesel engine, and it pushed a rapid-development program for Cane Cultivator-and-Plow tractors based on F-30-N narrow-tread models. Sugar cane had come back in Louisiana, and planters needed tractors with high ground clearance.

Suddenly, everything Johnston and Sperry supervised felt rushed. They never had enough time or personnel to do the work. Their efforts and IHC's product inventory expanded geometrically. Sugar-cane versions of F-30s rose high off the ground; Orchard, Industrial, and W-Series standard-tread variations of the Farmall F-12 were tested and went into production.

Then, for a few days, all work halted. Legge, the shrewd champion of tractor development, died of a heart attack in his garden as he pruned his lilacs on December 3, 1933. After the funeral, the McCormicks and the board of directors elected McKinstry president. He would serve only until New Year's Day 1935 before retiring.

During early 1934, the Naming Committee relabeled the improved gas-engine, wide-front 22-36 tractor as the WA-40 while the first diesels became the WD-40. By May, GPED completed seven WA and three WD prototypes. A newcomer to the EC and NWC meetings was Fowler McCormick. Born in 1898, Fowler was Cyrus Hall McCormick's grandson. A Princeton graduate, he drifted through several interests, including psychology, music, and accounting. In 1928, Legge suggested to Fowler that it was time to join "The Company." Starting in the apprentice program, Fowler learned manufacturing, engineering, and sales. In 1934, McKinstry named Fowler head of foreign sales.

In late May 1934, IHC found itself needing to repair its reputation. A combination of design, manufacturing, and field service problems produced an ill-fitting air filter for the diesel engines and the gas F-12s. Johnston explained: "The numerous complaints of excessive wear of the F-12 engines are largely due to dirt entering through inefficient air cleaners . . . It results in excessive wear of pistons, sleeves, rings, bearings, and crankshafts which in turn results in excessive oil and fuel consumption and loss of power." (Sperry defined acceptable oil use as 1 quart (1 L) per ten-hour field day after three hundred hours of use; these tractors used 1 gallon [3.8 L] a day.)

IHC's policy was to make it right by replacing faulty air cleaners with new ones, making engines airtight by, if necessary, thoroughly overhauling the engine and replacing worn parts. GPED designed new pistons with four rings instead of three. One source of the dirt was residual sand and metallic chips from Milwaukee and Tractor Works castings. McKinstry ordered them to install a filtering system for the lubricating oil they used to run in engines prior to installation. IHC had these systems in operation at the Fort Wayne gas-engine plant and at Farmall Works.

McKinstry kept records but released no totals for this repair. He had learned how to manage from Legge, and the new

president quickly found his voice, using it effectively to satisfy customers and to ask his managers to work more carefully and more wisely. It worked.

The pressure from Domestic Sales to provide a tractor for every farm and crop strained GPED by mid-1934. Although promoted to vice president of engineering, Johnston discovered that he couldn't slow the flood of new work. Two new projects lined up behind each one completed. Engineers returned from one test trip, filed reports, and left for another. The W-40s would start production even as 1934's Midwest drought slowed demand for all IHC products.

The F-30 engine displaced 284 cubic inches (4.654 cubic cm), compared with the F-20's (and Regular's) 220 cubic inches (3,605.2 cubic cm). Cylinder bore grew to 4.25 inches (10.8 cm) from 3.75 (9.5 cm) while stroke remained at 5 inches (12.7 cm).

F-30
McCORMICK-DEERING
FARMALL

When the costs for working at this pace came due, the price was high. At noon on June 27, 1934, McKinstry and Johnston shut down the 12-Series tractor production line. It was the only way to change parts before the 12-Series tractors left the plants. This delay permitted outside makers of new air filters and elements to deliver adequate supplies, so Tractor Works could remedy the problem before production resumed on July 9.

The entire repair program cost IHC $750,000. In four regions—Central, Southern, Southeast, and East—nineteen branches needed help; the thirty-six others around the country did not. The Service Department trained sales agents and sent them out to make repairs. Manufacturing estimated that perhaps 10,000 diesel, F-12, or crawler tractors needed service, ranging from simply tightening or replacing air filter canisters and elements to full top-to-bottom engine rebuilds, transmission repairs, and, in the case of crawlers, track replacements.

Johnston harped about keeping dirt out and oil in. IHC's sales organization, hungry for products and concerned with manufacturing costs, continually postponed his efforts to make tighter machines. In early August, Baker sent a new wide-front four-plow tractor, the CW-40, to Hinsdale. To try to avoid quality problems, Johnston asked for another year for testing and development. McKinstry told him to test harder and refused to delay production.

GPED staged final sign-off tests of the third preproduction W-40 Series in early October 1934 in Phoenix, Arizona. Johnston chose the desert to guarantee challenging conditions.

The only problems came in the transmission. While the transmission never failed in tests, the heat and dust taxed it. In Johnston's ideal world, GPED wanted to upgrade the transmission before the tractor grew from the 15-30 to the 22-36. No one envisioned the power of the diesel. Larger gears wouldn't fit; designing a new case and testing a transmission would add two to three more years. Again, McKinstry refused to delay introduction. Disputes such as these between Sales and Engineering set the stage for a drama that would play out over the next half century.

While the GPED was overworked, Benjamin was not resting. In early October 1934, at Hinsdale, he showed NWC and executive officers "a new means of attaching implements to the F-12 tractor," demonstrating both a No. 90 plow and a middle-buster. Predictably, the Sales staff and McCaffrey wanted all of it immediately.

Benjamin and Sperry sought to introduce this new hitch for plowing after the 1935 harvest. They could deliver F-12s earlier because hitch modifications were small. Implements were the problem; the list to be offered grew like a weed. With tractor production at 2,000 per month and the 1935 fall harvest ten months away, tractors and enough implements had to be in dealers' showrooms by July. Sales convinced everyone that tractors without implements were preferable to new implements without a tractor. This strategy gave them time to advertise to farmers.

With McKinstry's admonition about no delays searing their ears, Johnston and Sperry advised McCaffrey that their large six-cylinder diesel "with recent modifications could be run safely at a speed of 1,500 to 1,600 rpm, and could be depended upon to develop the horsepower required."

They hoped McCaffrey would support their efforts to build thoroughly tested, high-quality machines that would only be put into production when they were ready. McCaffrey was learning his executive skills from McKinstry. McCaffrey was not the last one who would disappoint the engineers in the decades to come.

McCORMICK-DEERING
FARMALL
F-12

FOUR

1935–1941

PROBLEMS (AND SUCCESSES) CONTINUE

The 1938 F-20. Perhaps it's the narrow wide-front end that's captivated them, but the herd came closer to inspect the machine. These weren't that rare, however. IHC produced more than 154,000 of its F-20 models.

When the F-12 first went into production, the tractor used a Waukesha-built engine designed by Fuller & Johnson Manufacturing Company because GPED's own was not yet ready. However, once IHC began using its own engines, problems quickly developed. The inaugural NWC meeting of 1935 presented a case of déjà vu: IHC had trouble again with piston ring wear, particularly with the oil control rings.

The 1936 F-12. After assembling twenty-five preproduction versions of this tractor, Rock Island Works got down to business and series manufacture started on January 11, 1933. IHC continued to produce these models into 1938.

This was just months after field service technicians had completed the "First Change."

In the case of the F-12, the kerosene manifold didn't heat the fuel well enough to vaporize it completely. Liquid flowed past piston rings, diluting lubricating oil. Worse, Leonard Sperry had sampled manifolds and found that in 40 percent of them, the cored hole missed the carburetor. Another cause that was out of IHC's control: bad fuel. Distributors, stuck with poor-quality heavy oil, added high-test gasoline. This blend broke down. Combustion did not consume some of the elements, leaving

damaging deposits. It was a geographic consideration. There were no problems in Texas, yet nearly every tractor sold in Little Rock, Arkansas, and more than half from Memphis, Tennessee, required an overhaul and a new crankshaft.

"Should we stop production of the kerosene-engine F-12 tractors?" Sperry asked. McCaffrey voted "no" on interrupting production. "We are practically current on orders," he said. Johnston and Sperry hurriedly created new manifolds and revised carburetors with a smaller venturi and a new fuel nozzle. They fitted these "Second Change" kits to tractors still on the assembly line.

In early February, GPED's managers examined the new hydraulic power-lift built into the F-12 Farmall, which was driven by the PTO shaft. They compared it to an implement lift from a John Deere Model A tractor. Deere's system provided farmers with hydraulic lift and gravity drop. IHC patent lawyers worried whether its proposed system placed "forward of the rear axle" differed enough from Deere's placement "at the rear of the tractor."

Meanwhile, EC members watched the new prototype F-20X, built along F-12 lines. This was Johnston's newest application of the semi-unit-frame design of his 1914 Mogul 20-40. Sperry presented Sales with a new advantage that F-12s offered over John Deere models: GPED's power implement lift put pressure on the implement to set it into the ground after it was dropped.

Another F-12 invention that eventually spread to the rest of IHC tractor lines was the Quick Detachable "QD" Drawbar. By April 1935, GPED had working prototypes. Manufacturing accelerated the pace of all tractor lines to develop a steady schedule that prevented periods of plant inactivity, but this action caused the stockpiling of unsold tractors instead.

"When tractors stand exposed to the weather more than thirty days," A. W. Seacord of Domestic Sales warned NWC members, "it is usually necessary to repaint the tractor at a cost of about $5.00 each. This occurs whether the exposure is in storage at factory yards or outside at branches or with dealers." GPED proposed using new "synthetic paints which must be sprayed and which require a somewhat different provision in drying ovens . . . [and] will allow an exposure of four to six months. If the tractor is shipped or sold during that time it can be wiped off with an oily rag and the appearance will be substantially that of a new tractor." The company soon adopted the new paint technology.

At the May NWC meeting, Fowler McCormick, as foreign sales manager, joined the discussion on reestablishing tractor manufacturing in Europe. Their prime candidate was the F-12. By January 1936, Sperry updated the NWC about European matters. He had learned German was using *Treibstoff*, a combination of gasoline and alcohol "of better quality than diesel fuel," which would work well in the F-12's gas/kerosene engine.

FARMALL

The growing adoption of pneumatic rubber tires forced GPED to reconsider tractor speeds. Previously, top gear speeds had developed haphazardly. Now IHC needed continuity across product lines when each agricultural tractor was tested at the University of Nebraska. Baker of Engineering reminded his colleagues that it was necessary also to have adequate brakes for increased speed: "Such brakes would soon be made necessary by legislation."

The first of five preproduction F-21s (the F-20X) reached Phoenix, Arizona, in early April. Two remained in Chicago for Deering Works to design implements for their hydraulic lift mechanism. Early reports were very favorable. One F-21 in Phoenix ran 145 hours with a three-bottom plow working 10 inches (25.4 cm) deep. With GPED and NWC both watching tractor weights, Sperry explained why the F-21 weighed 400 pounds (181 kg) more than production F-20s.

The new tractor had 54-inch (137 cm) wheels with heavier spoked rims and an extended axle. The model's wheelbase was 6 inches (15.2 cm) longer than the F-20 to accommodate the front cultivators, while the transmission weight came from the need for a 15 mph (24 kph) speed on pneumatic tires. The F-21 engine weighed 958.5 pounds (434.8 kg), just 2 pounds (0.9 kg) more than the F-20 power plant, and produced nearly 4 more horsepower.

"If we are on the wrong track as to the type of tractor," Sperry said, "we must redesign it. There is nothing much we can do to reduce the weight of the tractor now proposed for the purposes for which it is designed." Field tests showed that the John Deere Model A gained one round on the Farmall in every ten rounds plowed. While the F-21 was nearly 900 pounds (408.2

kg) heavier than the Deere, Sperry argued that its higher speed, quick implement-mounting system, now referred to as Quick Attach (QA), and improved plowing power moved it past the competition.

If they could keep too many more new ideas from infecting their designs and slowing development, Johnston and Sperry both believed 1937 would be a strong product year. They had a new Farmall and two new crawlers. By early August, Sales pressed GPED to adapt adjustable rear tread to the F-30 as well as the F-21, now referred to as the New F-20. In mid-October, Sales proposed that the GPED should redo the big Farmall along the lines of the Intermediate New F-20, or F-22. Engineering

also began work on a new tractor to replace the W-40. Sales had considered the W-40 a temporary tractor, even referring to it as the "converted 22-36 tractor."

In 1935, work progressed at the Experimental Department on synthetic enamel paint technology. The question was whether to change from gray varnish to gray synthetic enamel paint, as suggested by Seacord, or Sale's desire for something new. Various colors were discussed. The industry already had enough green tractors between Deere, Oliver Hart-Parr, and some of their own lines. Case and Ford seemed devoted to gray. The GPED wanted to make the tractors more visible on roads. Red continually emerged as the answer. By summer 1936, tests showed the new red synthetic enamel held up to steady exposure to sun and elements better than the previous gray had. The decision was made. On November 1, 1936, the first 1937 model tractors rolled out of Tractor Works and Farmall Works wearing "Harvester No. 50 motor red synthetic enamel paint." Wheels remained dipped in Harvester red varnish.

Early in 1937, R. M. McCroskey experimented with high-test gasoline in F-20 tractors using high-compression pistons. Johnston objected, pointing out that "our tractors, including the F-20 Farmall, were designed with an 'engineering balance' as to power and strength throughout their various parts and it would be foolish to put more power in the engine than the chassis or power [train] could take care of." This concern haunted IHC through the remainder of the corporation's life. Still, McCroskey stressed the importance of preparing to use the higher-performance fuel and the benefit of its improved fuel economy despite its higher costs.

"Farmers want low cost of operation," McCaffrey argued, "which we have shown is accomplished by our tractors using low-cost fuels. High test is not low cost." The question of performance arose again when Sperry told NWC members that nearly all tractor manufacturers that submitted machines to the University of Nebraska ran the tests on distillate, which was low-cost fuel, available throughout the United States and Canada.

"With the proper compression," Sperry explained, "this [fuel] gave in the neighborhood of 4 percent better horsepower than a similar tractor designed for kerosene and 15 percent better results in fuel economy."

W. E. Payton, service manager at the St. Louis branch, sent Sperry a telegram on April 8, 1938, telling him that two

above: **With 220 cubic inches behind the bright red paint, and IHC's own E4A magneto providing ignition spark, Ed Johnston's engineers pulled 15.98 drawbar horsepower and 23.8 horsepower off the belt pulley in early versions of the F-20.**

opposite: **IHC began production of the Increased Power F-20 in early September 1937, to keep pace with competitors in an ever-expanding horsepower race.**

left: IHC's W-30 models first appeared in 1932 while its "official" orchard model, the O-12, debuted in 1934 as a contemporary of these bigger, more powerful Orchard California Special models.

right: Original equipment and even the decals held up well in a sympathetic California environment. It's apparent from details such as this that while the tractor saw use, it was not abused.

opposite: The 1938 W-30 Orchard California Special. Well-known IHC collector Mike Androvich likes time capsules. These are unrestored machines that remain completely original in their equipment and their appearance. For restorers, these are great resources to see how the factory assembled them or how early operators modified them for better service.

prototype Allis-Chalmers Model B tractors worked fields nearby. Sperry dispatched McCroskey. He and Frank Bonnes from Domestic Tractor Sales were impressed with the semi-unitized-frame and torque-tube construction. Bonnes was perhaps more affected by the area and its small scale of farming. J. M. Strasser, assistant branch manager and his guide, told Bonnes that around St. Louis, there were more than 1,200 farms of between 5 and 40 acres (2 and 16.2 ha), farms too small to use any tractor other than a one-row machine.

"We have, I believe, just missed this market with our 12-Series tractor," Bonnes alerted the NWC. "The price of our F-12 has gone steadily upward until at the present time these small garden farmers will not make the investment required. . . . If this [Allis-Chalmers] tractor . . . can stand up, this little tractor is certainly the greatest threat to our F-12 business today."

IHC powered tractors of this era with its 3.75x5-inch inline four-cylinder engine with 220 cubic inches displacement. In tests at the University of Nebraska, the F-20s developed a maximum 26.7 horsepower on the belt pulley and 19.6 off the drawbar running on inexpensive distillate fuel.

Through a stringent weight loss program, GPED reduced the F-22 weight to less than the F-20. However, under certain farming and implement use conditions, the tractor front end was too light. Johnston added 400 pounds (181.4 kg) back onto the F-22.

On March 17, 1938, McCaffrey urged quick decisions so F-22 production might begin on schedule on November 1. Manufacturing had promised him 4,000 tractors available by January 15, 1939. Tooling and raw materials costs were rising, but salesmen were confident. Then, on April 13, revised manufacturing cost estimates made McCaffrey reverse himself.

"The F-20 is one of the best tractors in the Harvester line," he said. "The F-20 is still salable if certain improvements are put into it. It will cost $2.5 million to put the F-22 into production and it will have to sell for $1,025 against $985 for the F-20, both on steel. This higher price will put us out of the market."

McCaffrey then railed against the streamlined sheet-metal treatment on the new prototypes that had been created by famed industrial designer Raymond Loewy. "And this 'dressing up' of the tractor!" he said. "A farmer puts a tractor out into the field to work. This dressing up does not mean anything. A tractor is built to do a job quickly and cheaply. Too much has been done on the appearance factor."

GPED design engineer A. W. Scarratt was incredulous. "This machine [F-22] has all the things that Sales [has] asked for: clearance under the machine, adjustable tread, new transmission to offer speeds from 2½ to 21 miles per hour [4 to 33.8 kph], provision for pneumatic tires, variable speed governor, good brakes, better steering wheel, water pump, foot accelerator, a standing platform, and it has the increased power everyone asked for,"

he said. Further, "the effect of style consciousness crept in and so we dressed up the job. The fact of our having 'styled' this job has caused no penalty in the tractor mechanically and this sheet metal can be taken off."

"But our competitors do not have all the improvements in one machine that we have in the F-22," McCaffrey replied. Sales felt wary of being first with so many new features.

At its April 27 meeting at Tractor Works, the NWC killed the F-22, creating instead a new three-plow tractor, the F-32. The NWC endorsed a one-, two-, and three-plow tractor, but with new designations to avoid mushrooming confusions with Farmall F-numbers. The one-plow F-10, authorized in September 1937, was renamed the 1-F. The two-plow F-15 with a new 22-horse-power engine became the 2-F. The F-32, authorized back in April 1936, became the 33-horsepower, three-plow 3-F. Johnston announced that he would have manufacture-ready prototypes in July 1939, "[t]his date being dependent on Sales not changing

their minds again as to what they want in sizes and powers," he added. The meeting minutes did not reflect the tone of his voice.

For the F-10/1-F, Johnston proposed an L-head engine and a new overhead-valve version. The F-15/2-F provided drawbar performance to match Allis-Chalmers' WC. McCaffrey said that the F-22, at 29 gross horsepower, was not enough to call it a three-plow tractor. "A three-plow tractor," he explained, "should have at least 31 and, better, 32 horsepower."

Johnston reminded him that the 33 drawbar horsepower F-32/3-F would weigh perhaps 200 pounds (90.7 kg) more than the F-20. However, few parts of the F-22 could be used on the now-proposed F-32/3-F. Charles Morrison wondered aloud, "Under the present plan, then, does the F-30 just fade out of the picture? And what about a four-plow model?"

Johnston described to them two tractors known in GPED as the F-40 and the W-42. "The engine proposed [a new 5x6.125 four-cylinder with dry liners] was in two forms, one with the

engine base as part of the tractor frame and one with the engine to be mounted on a more conventional chassis structure. The W-42 would be a real four-plow tractor."

"We must have a four-plow tractor," McCaffrey concluded.

While farmers demanded larger implements and Benjamin's implement makers responded, Johnston's tractor engineers encountered problems with the power lift mechanisms. Scarratt and Benjamin used 5-inch (12.7 cm) diameter hydraulic cylinders (replacing previous 3-inch [7.6 cm] sizes), mostly to accommodate heavier F-30 equipment. Carl Mott designed a system that lifted left or right gangs or front or rear sets independently. A double-acting cylinder valve-and-spring mechanism raised and lowered front and rear equipment in sequence.

GPED discarded the overhead-valve engines in F-10/1-F prototypes. Sperry cited large tooling expenditures as the primary reason. NWC authorized two more prototypes in late June 1938, one using an L-head engine and the other with a redesigned F-12 power plant for comparison purposes. Sales kept pressure on Sperry to complete testing in time to launch production on March 1, 1939. They calculated that if GPED followed customary preproduction programs, including several-month lag times between testing, prototype costs, and design changes, then "our competitors will have sold 40,000 tractors before we even enter the market." Then H. D. MacDonald, from Sales, confused things. While pushing an accelerated testing program, he also insisted on adjustable front and rear axles to straddle two rows.

The NWC and EC met at Hinsdale on August 2, 1938, to watch several demonstrations, including a mock-up of a 2-F tractor plowing against an Allis-Chalmers WC, both using two-bottom 14-inch (35.6 cm) Little Genius plows. The participants began to second-guess the prototype.

McAllister asked, "Are we on the ragged edge as to horsepower? We must consider having ample power for two plows and also we must consider cost!"

Johnston, frustrated by company management that wanted heroic efforts in development and miracles in meeting deadlines while juggling too many projects in the air, snapped, "If more horsepower is wanted, we must start all over. IHC's outside industrial designer [Loewy] has approved the proposed

The 1939 F-14. At the other end of the spectrum from unrestored original tractors is this sparklingly restored jewel. Michigan collectors George Morrison and Bob Findling labor long and hard to get their machines to look this fine.

1-F tractors. That was before the NWC saw the Allis-Chalmers Model B with adjustable-track widths from 40 inches to 52 (101.6 cm to 132.1). Allis sold nearly 11,000 tractors by year-end.

Fortunately, Baker already had an adjustable front and rear axle. McCroskey now reasoned that perhaps 90 percent of the 1-Fs would be sold with these axles. When Johnston showed a prototype row-crop 1-F, with a single front wheel to accommodate two-row planters or cultivators, McCroskey feared that purchasers would demand for it to use larger implements, which would require adding more power, strength, and weight to the machine. McAllister and Morrison agreed, telling Johnston to continue experiments to perfect "a three-wheel type 1-F, but it must not be offered to the trade unless it is decided to do so later by force of circumstances."

By "circumstances" they meant the pressure of competition. During a June 13, 1938, conference on "Development of New Machines," McCaffrey set out six guidelines for information he required before Sales would consider new development or major redesign projects. They were:

1. Competitors' weights
2. Competitors' list prices
3. Our weights required
4. Our required list price
5. Our product cost necessary to establish this list price
6. Estimated sales

Projects stood a better chance of a Sales Department endorsement if someone else already made it at a price IHC could meet or beat. Fowler McCormick gained influence here: the

styling of the tractor, though it is not just what he would like to have."

"It would be very desirable," McCaffrey mentioned at the end of that meeting, "to announce all three Farmall tractors to the trade at the same time." The board settled on phase-out plans for F-30, F-20, and F-14 models; new Farmalls would require factory production space, and markets for older tractors would surely end after introduction in September 1939. On October 3, 1938, they agreed to drop the W-30 and 10/20 tractors before May 1939, to meet factory capacity.

During a November 9th conference at Hinsdale farm, another McCaffrey issue returned to haunt NWC members. The previous March, he had argued that adjustable-tread tractors made up only 20 percent of the market and were not needed on

quintessential farmall

tHE FARMALL H was introduced in 1939 as a replacement for the F-20 and was the most popular Farmall. The H was one of the first products of industrial designer Raymond Loewy's genius in restyling the entire International Harvester line. Loewy, it will be remembered, was responsible for the rakish Studebaker car styling of 1953 as well as for the famous Air Force One paint scheme first seen in 1962. Loewy's smooth contours and bright red sheet metal make even a 1939 Farmall look completely modern more than eighty years later. Production continued until 1953, with almost 400,000 units sold.

Developed from the F-20, the Farmall H had a modern, higher speed engine with water pump cooling and a five-speed transmission. A starter and electrical system were optional. Gasoline and distillate were alternate fuels. The H, significantly, was the last tractor to be tested on distillate fuel at the University of Nebraska.

The frame layout allowed the mounting of cultivators and other implements that would also fit on the H's big brother, the Farmall M. While the dual tricycle front end was standard, a wide front was optional. High clearance, or HV, models were also available.

The Farmall Super H, which came out in 1953, was much the same as the H, except for increased displacement giving about 25 percent more horsepower. The Super H also had disk brakes and, later in the model run, live hydraulics. Super H production ended in late 1954. A few less than 22,000 were delivered.

— *Robert N. Pripps*

The 1939 Model H. IHC manufactured its first Model H tractor on July 3, 1939. It quickly became the best-selling Farmall. Farmall Works produced more than 10,000 in 1939 alone. It would share a wheelbase with the larger Model M so that Bert Benjamin's implements functioned interchangeably. This meant that farmers didn't have to purchase two sets of tools if they owned both an M and an H.

board named him vice president of Manufacturing in 1938. His staff determined costs that influenced IHC's retail prices.

Throughout that winter, GPED continued working and then IHC launched a number of new tractors. On May 23, 1939, Baker signed off on production orders to start assembly of 1-F tractors, now called the Farmall A. The first machine rolled out on June 21, 1939. On June 6, Baker approved the new 2-F as a Farmall H and the 3-F Farmall as the M. Model H and M assembly began July 3. Implements designed for the three lines went on sale on July 20. Baker released the Farmall B on August 8 and the first one emerged on September 5.

The following year, on August 12, Baker signed off on a Narrow Tread Farmall B, providing rear tread widths adjustable from 56 inches to 84 (142.2 cm to 213.4), in 4-inch (10.2 cm) increments. Manufacture began October 15. Nine days later he launched production of a High Clearance Model A, the AV. IHC manufactured the first one on January 10, 1941. For the time being, the Sales Department was satisfied.

left: The 1939 Model M with Elwood Front-Wheel Drive. Elwood Manufacturing began producing 4WD kits in the mid-1950s. Farmers needed increasingly efficient methods and equipment to get engine power onto the ground. More power to the rear wheels alone sometimes resulted in wheel spin.

right: The AV's offset seating position allowed adult operators to work comfortably on this compact tractor. It also permitted excellent crop visibility during cultivating.

INTERNATIONAL
INTERNATIONAL HARVESTER
FAIRWAY 14

above: Fairway 14 models originally appeared on wide steel wheels with "sod-puncher" lugs—small, rounded cones that aerated the lawn as the tractor drove over it. Many operators converted to pneumatic rubber.

opposite: The 1939 Fairway 14. International assembled just 114 of these higher-power Fairway models throughout 1938 and 1939. These compact models measured just 105.5 inches long overall and stood only 50 inches high.

McCORMICK
FARMALL
A

FIVE

1941–1944

INNOVATION AND PRODUCTION SURGE

The 1941 Model A. IHC produced more than 117,000 of these small farm tractors between 1939 and 1948. In 1940, the base machine sold for $515.

The NWC pushed along one large engineering project after another throughout 1941. Most involved diesel engines. During a Farm Tractor and Implement Group (FTIG) meeting on February 27, Sperry, Morrison, McCaffrey (just named vice president of Worldwide Sales), and soon-to-be-elected IHC president Fowler McCormick (succeeding his father, Harold) all agreed to build a sample Model H high-clearance tractor with a diesel engine.

However, the IHC's workforce labor was not happy. The day after the diesel high crop approval, some 6,500 workers struck McCormick Works. By March 3, nearly 15,000 employees had sought representation by the Farm Equipment Workers (FEW) organizing committee. After reaching an agreement, strikers returned to work on March 23.

In late September, the GPED members discussed the Farmall B straight-axle tractor and a new Farmall E model that incorporated live hydraulic implement lift and independent PTO. Because of the difficulties in updating the B's older technology, the NWC decided to go ahead with the E, incorporating a straight axle. Tractor Division Experimental Engineering continued peacetime developments even as factory space was devoted to armament production.

After the Japanese attack on Pearl Harbor on December 7, 1941, World War II became an American reality. With only a five-week supply of pneumatic tires, the IHC discussed shipping new tractors on steel wheels because the government curtailed rubber deliveries. Worried that steel-wheel technology was no longer adequate for more powerful tractors, GPED devised stronger wheels and recommended 5 mph (8 kph) transport speed limits.

On January 2, 1942, Scarratt showed the NWC a wooden full-scale model V-8 engine developed at Fort Wayne, Indiana. GPED designed the V-8 in two displacement sizes, one capable of a maximum of 460 cubic inches (7,538 cubic cm) and a larger one up to 655 (10,734). "The desirability of compactness in tractor applications," he explained, "is nearly as vital as in truck usage."

GPED shelved the Farmall B Straight Axle tractor on January 16, 1942, due to wartime demands for raw materials as well as needs for the Farmall E's development. The war sidetracked QA hitch development and the QD Quick Detach system. The straight-axle B eventually would reappear in 1947 as the postwar Model C.

Arnold E. W. Johnson offered a GPED innovation on March 25: "New machines must be designed to fit the crops and farming practice of numerous individual localities. Direct-mounted implements on basic front and rear frames may improve this. To each we can attach a variety of working tools [that] can be removed intact with the adjustments preserved, saving the operator time required for resetting them. . . . Hydraulic fingertip depth control [and] power-lift eliminates varieties of hand lever assemblies."

Engineering christened this system the "Frame-All." Unlike the Ford-Ferguson three-point rear hitch with one cylinder and one control lever operating implements at the rear of the tractor only, Frame-Alls used two double-acting cylinders and two levers, allowing adjustments on one side of the tractor or the other, front or rear. It could accomplish delayed lift and drop of front and rear implements.

Two days after the Frame-All Model H introduction, McCormick clamped tight security on it and the QA system with hydraulic controls. He theorized that "after the war, farmers would have used up equipment and would need something new. Sales would want something new to sell."

Scarratt suggested replacing A and B models with this E straight-axle tractor (the Model C). "Adaptations of the 'QA-Frame-All' and hydraulic fingertip lift control should be confined to Farmall E," he argued, "because the basic principles of this scheme of tools and attachments lends itself readily to this tractor which is more similar to the H and M in general outline than either the A or B tractors."

War production consumed most of IHC's factory capacity. But once tooling was set up, GPED experimental engineers felt little pressure to develop new machines every ninety days. They had the time to perfect the Frame-All and the E. By late fall 1942, GPED had completed Scarratt's first Farmall E, and they started assembly of a "Reverse Direction Super Farmall E."

T. B. Hale, in regional tractor and implement sales, addressed a dealer's meeting in Dallas on March 23, 1943. It had taken IHC five years since April 1938, when Bonnes visited the

above: **The 1941 Model AV. IHC data lists the higher-clearance Model A tractors as 9 inches (22.9 cm) longer overall and nearly 400 pounds (181.4 kg) heavier than the standard-clearance Model A tractors. The company produced 3,603 of the AV models.**

opposite: **The seating and steering offset is apparent in this front view. Adjustable front and rear tread width made this compact tractor versatile and valuable to small-farm operators.**

St. Louis area, to form a product plan for mid-America's smaller farms. Baker joined Sales and Manufacturing managers to host meetings from New York City to Los Angeles. The executives questioned 124 regional branch managers and dealers about IHC's current equipment and its uses, the competition and its advantages, field service and factory changes, mistakes of the past, and rumors about the future.

"The consensus was that many wartime developments might well be carried over into the design, production, and uses in commercial industrial power industry after the war," Baker told Johnston and McCormick. "Dealers leaned toward expanding the line upward in power and they expressed need for diesel power units as high as 200 horsepower."

In July 1943, Sperry distributed an eight-page "Survey of Potential Demand for Farmall 'X' Tractors (smaller than the Farmall A)." The report noted the Farmall A was appropriate for farms from 40 to 70 acres (16.2 to 28.3 ha). However, the 1940 U.S. Census revealed that of the 5.7 million farms reporting crop acreage, 3.3 million (or 58 percent) were smaller than 40 crop acres (16.2 ha). Of those, 2.2 million had an annual gross farm income of $400 or more (about $6,765 today).

Sperry wrote: "In the fifteen years of Farmall tractor type selling, IHC has sold 733,000 Farmalls." The potential, therefore, was "to reach the untouched market demanding smaller equipment." The Farmall X would "do the work of two or three horses or mules, be a four-wheel tractor, row-crop type, with 8 horsepower on drawbar. It would be designed 'CultiVision-style' [engine offset to the operator for better crop visibility] for QA Quick Attach machines for truck garden and field work whose retail price is not to exceed $400." They proposed a complete line of implements, tools, and attachments. To keep manufacturing costs at around $213, necessary to meet a $400 list price, Baker and Sperry proposed a new two-cylinder, parallel, upright engine for the X.

FTIG released specifications for the Model E/C. It ran the basic Model B engine at 1,650 rpm, with pump-circulation engine cooling. The E/C chassis used a straight axle with 36-inch (91 cm) rear tires, 15-inch (38.1 cm) fronts, 21-inch (53.3 cm) ground clearance, and adjustable tread from 48 to 84 inches (122 to 213.4 cm) on an 86-inch (218.4 cm) wheelbase. The transmission provided five forward and five reverse speeds. The E/C offered continuous running PTO and hydraulic lift pump with Touch Control. It added the simplified and improved QA implement mount system and had "styled and non-vibrating sheet metal enclosures, over the engine and rear fenders with cutouts for implement movement."

On September 1, 1943, McCormick's engineers attended to a full range of tractor products for agriculture, industry, and construction. McCormick saw stressed managers pulled taut by the workload and variety of projects they had to manage. Like his grandfather, he recognized people's strengths and weaknesses. IHC had strengths, too, but its tendency to force division general managers to do too much, he feared, could harm the company. In meetings he had heard lapses and errors.

With millions of dollars at risk in development and tooling costs, McCormick knew mistakes could devastate a corporation's

McCORMICK
FARMALL
IH
H

The 1944 Model M. IHC's new Model M was the next tractor to get Touch Control development. GPED had this prototype, with its experimental hydraulic lift levers, photographed on February 5, 1944. *State Historical Society of Wisconsin, WHi (X3) 52007*

health. He conceived a plan to reorganize IHC where areas of expertise and interest influenced a manager's job selection. He separated Farm Tractors, Farm Implements, Industrial Power, Motor Truck Division, and the general line, including refrigeration and other products, from the tightly centralized rule his grandfather had established. He wanted first to reduce executive workload while giving autonomy to division vice presidents. Second, he intended to diversify the EC, which was top-heavy with eight former sales managers as members but only one engineer and one manufacturer.

At the new Tractor Division, testing the Intermediate H and M tractors with Frame-All hydraulic controls revealed that the larger twin hydraulic cylinders were inadequate for middle busters or four-row cultivators. Operators found control levers poorly placed and they needed to hold onto them until the lift or drop was complete. As development wore on, E. F. Schneider from Sales grew impatient with discussions of further testing. "In recent years competitors have put new machines into production after a test with only one experimental machine. We should release our new machines, whenever possible, without the usual preproduction lots, in order to be in the lead rather than to follow others." Industry-wide, manufacturers began to rely on customers to complete their final development programs.

Just before Christmas 1944, on December 20, in the Motor Truck Sales Room on the southwest side of Chicago, FTIG showed off the Farmall X prototypes to nearly four dozen managers from Engineering, Sales, Manufacturing, Service, and Executive committees. Scarratt played master of ceremonies, introducing the Farmall X, which they sometimes referred to as the "Baby Farmall."

McCaffrey's Frame-All program was to add twenty-five A and B tractors to the tests. It nearly brought manufacturing to a halt on January 2, 1944. A and B assembly-line production was underway. Manufacturing had orders for 5,000 tractors through

left: **Keith Feldman dug into the history of this roller. He learned that Jacobs Farm Equipment, Ltd., of Essex, Ontario, produced just five of these rollers on Farmall H tractors.**

opposite: **H model tractors manufactured during wartime eliminated the top speed transport gear. IHC also reconfigured fourth gear, slowing it from 7 mph (11.3 kph) down to 5.375 mph (8.7 kph) for tractors on steel wheels.**

1944. Interrupting this to hand-assemble twelve prototypes was impossible. Baker understood the need to complete Frame-All hydraulic control system tests but felt "Manufacturing should cooperate fully to become familiar with this important new development." McCaffrey, just elected to the board of directors as an IHC second vice president, had aimed at July 1, 1944, as the production approval date, based on Frame-All models testing nonstop through May and June. Now, it appeared he would have to reset this to July 1, 1945, unless major effort moved the program.

Manufacturing was adamant. It would not interrupt two assembly lines for a single A high crop and four BNs. McCormick agreed to slip introduction back to 1945, but he reminded manufacturing vice president H. K. Kicherer that Canton Works had a variety of twenty-five Frame-All implements in A and B tractor sizes that needed thorough testing as well.

McCormick then ordered Kicherer to "make an immediate, intensive study of the hydraulic system and gain the necessary knowledge of, and experience with, this unit preparatory to its production."

The Farmall E/C program, launched in September 1939, was in jeopardy in early January 1944. FTIG had converted a standard drop-axle Farmall B into a straight-axle tractor. This required costly revisions when testing revealed strength problems and implement mounting constraints. As a result, in late September 1941, FTIG had created the new straight-axle Farmall E/C prototype that it demonstrated in mid-August 1943. Now, getting cold feet, the EC ordered FTIG to greatly reduce cost and weight, and eliminate "certain features it felt [it] could not afford to incorporate in a tractor of this size and capacity."

"The Farmall E," Baker reminded FTIG members, "was the pilot model for a complete line, because it would be restyled and incorporate features which it was felt were essential for an up-to-date future line of tractors. Such a tractor would become the forerunner of a new line. The M size has sufficient horsepower to satisfy power requirements for a complete line of attaching tools which are considered in new implement developments, including

AMPERES
10 0 10 20
DISCHARGE   CHARGE
IH
364458-R91
O
D
B

In IHC's plants, workers busy making half-tracks and torpedoes for the war effort found it difficult to remember farming as usual.

[mounted] harvester threshers." The final coffin nail, however, came from Archer in Sales.

"Because the Farmall E was larger than the A and B, but smaller than the H and M, [it] would render both larger and smaller tractors obsolete prematurely, thus interfering with the sale of these tractors pending new developments." This was crys-tal-ball gazing, similar to what Sales had done worrying about McCormick-Deering 10-20 models after introducing the Farmall.

The war drained away time and energy; existing programs soldiered on. Just as during World War I, while the male popula-tion fought, wives and daughters ran the farms to feed and clothe the world. IHC's branches organized tractor operation and repair schools for women. The EC told Chicago it needed to adapt current production hydraulic implement lifts to existing F-20 and F-30 tractors. This was not easy. FTIG assembled one and found it required forty-two new pieces, including a fabricated angle steel frame. The chance for leaks was great, and GPED did not encourage the idea. However, because it would aid farmers at home, primarily women, GPED agreed to make it work cleanly and quickly.

FTIG had Frame-All A and B prototypes in fields by May 1944, but the EC decided to lease or loan the tractors "for test purposes only" and not sell them outright. They feared another

farmer falling in love with a unique prototype, as had happened with one of Johnston's International 8-16 four-wheel-drive pro-totypes two decades earlier. Leasing placed a financial burden on the farmer, who was more likely to use the tractor fully and report honestly any complaints and failures if it directly affected farm finances.

In IHC's plants, workers busy making half-tracks and torpe-does for the war effort found it difficult to remember farming as usual. Throughout Europe and the Pacific, soldiers and sailors used IHC products and those from other U.S. manufacturers in the fight for the return to living as usual. On June 6, 1944, D-Day, tens of thousands of Allied troops landed on the northern coast of France. Within two weeks, Allied soldiers captured more than 30,000 German soldiers occupying areas near the coast. The war was not over but progress was measurable.

One of the crown jewels of Wisconsin implement dealer Arden Baseman's Farmall collection is this Model A prototype. While rumors persist that another one has surfaced, this is the only Frame-All known to exist.

On July 11, tests confirmed success of the Farmall A single-cylinder hydraulic Touch Control, and B, E/C, H, and M double-cylinder systems. Questions of manufacturing cost and retail sale price rose against Manufacturing's production schedule. The start date slipped back from July 1, 1945, to September 1, after FTIG learned they should not use, advertise, or list the term "Touch Control" relating to its hydraulic system until IHC had it trademark-protected.

Archer from Sales met with McCroskey and Sperry from Engineering to schedule tests for both Model E/C and X tractors. Complete Farmall A Frame-All tractors weighed 1,856 pounds (842 kg); the new C weighed 2,150 pounds (975.2 kg), and Xs were only 1,058 pounds (480 kg). After calculating sales and cost benefits of the Frame-All tractors compared to standards, McCroskey reported that creating the Frame-All used new parts costing $47.01, while removing others worth $44.13. This yielded a net increase of $3.12 per unit in production costs. While this represented a small price, at this point, the EC envisioned a production run between 70,000 and 100,000 A, C, and X tractors.

On Thursday, September 14, FTIG engineers staged a massive show-and-tell for EC members, division general managers, and others in Tractor and Implement Divisions. Starting at 8:30 a.m., FTIG demonstrated what they believed were "sign-off" versions of Frame-All-equipped hydraulic Touch Control–operated Farmall A, B, C, H, and M tractors. There were no failures, no miscues, and no disappointments.

As Implement Group engineers parked the last demonstrator in the Hinsdale farm shed at 4:30 p.m., company cars and buses loaded up and rolled out of the yard. IHC management knew it had seen its future that day. IHC would have innovations, techniques, products, and tools to sell farmers who came home from the war.

big brother

tHE BIG M was the envy of farmers with lesser tractors, but the $1,200 price tag of the M was $200 more than that of its smaller brother, the Farmall H; in the early 1940s, that was a big difference. The M replaced the F-30 in the International Harvester stable in 1939 and, like others in the line, it featured Raymond Loewy styling. The new M offered a 248-ci overhead-valve four-cylinder engine. Either low or high compression heads were available for distillate or gasoline fuel. A five-speed transmission was used. The "Lift-All" hydraulic system was an option. The M, rated for three 14-inch (35.6 cm) plow bottoms, was available with a wide front end or with the dual tricycle setup. High-crop versions were also available with wide or single wheel fronts.

It was a surprise in 1941 when a diesel engine version of the M was introduced, called the MD. International Harvester had pioneered diesel wheel tractors when the WD-40 was introduced in 1934, but the Farmall MD was the first diesel row-crop farm tractor. Large-acreage farmers found the MD paid for its initial cost over a conventional M in less than a year because it used about a third less fuel, and diesel in those days cost about half as much as gasoline. Fuel consumption savings were even greater over gasoline and distillate at part loads.

The M-TA (Torque Amplifier) option was added in 1953. It was a two-speed powershift gearbox that doubled the gear ratios available.

The Farmall M was the first tractor to be assembled in Harvester's U.K. plant in Doncaster, Yorkshire, England. Production of the BM, or British Model M, began in 1949, some ten years after the M was introduced in the United States. The Farmall BMD diesel came out in 1952. — *Robert N. Pripps*

The 1945 Model M. It was big, powerful, and handsome. IHC hired outside industrial designer Raymond Loewy to "style" its new Letter Series models that first appeared in 1939.

Loewy's prototype designs displayed full engine covers trimmed with slender chrome strips. The Executive Committee vetoed the chrome because of costs, and engineers eliminated the covers to improve engine cooling, as seen on this and all subsequent production Ms.

SIX

1945–1954

CHANGING FOCUS, BLURRING VISION

The 1953 Super B-MD. While IHC produced just about 5,200 of these in the United States, records estimate the British M-diesel production at more than 900. Other than country of origin, however, the machines are very similar.

After manufacturing a diversity of wartime products, Fowler McCormick believed International Harvester could do more than be a farm equipment and truck maker. IHC had nearly 100,000 workers in seventeen plants, with annual sales of more than $100 million. The company had money to expand product lines and to buy new factories.

McCORMICK
FARMALL CUB

The Tractor Division staked its development money on a compact machine for the small two- or three-horse farm. Here the Naming Committee broke form. The Letter Series Farmalls began in 1939 with the Model Ms. Everyone had referred to the prototype as the Farmall X or the F. In September 1945, public relations man Art Seyfarth, patent attorney Paul Pippel, engineer Sperry, and the five other committee members named it the Cub. IHC's Sales Department aimed 45 percent of total Cub production at the East and Southeast regions.

Production began in Louisville, Kentucky, late in 1947. Nearly 135,000 Cubs rolled out over the next four years. Half the purchasers were first-time tractor buyers replacing horses

left: The 1945 Model BN. Chicago Tractor Works manufactured the first BN, or Model B Narrow (single) front-wheel tractor, on October 31. The BN provided a narrower rear tread width than the standard B offered. It could be as slim as 56 inches (142.2 cm). This catered to vegetable growers and truck farm operators.

right: The 1945 Model B. This was a typical Model B configuration with its cambered front wheels creating a small row-crop tractor. The B weighed 1,830 pounds (830 kg), about half of the Model H's 3,725 pounds (1,690 kg). These compact machines could spread out rear tread width to as much as 92 inches (233.7 cm). IHC manufactured a total of 75,241 of the Model B tractors.

opposite: The 1951 Farmall Cub. Conceived and developed as the Farmall X or Baby Farmall, IHC introduced the Cub in 1947. The Sales division had concluded that southeastern cotton and tobacco farmers could use a machine that was smaller than the Farmall A. The compact Cub proved a real working machine for small-acreage farmers. Of course, this philosophy made it the perfect tractor for large-acreage estates with lush gardens.

The 1947 McCormick-Deering Farmall Cub. As part of the Farmall Cub launch, dealers around the country staged events to introduce the compact tractor to potential customers. Three years later, in 1950, the Louisville plant would resort to white paint to get customer attention. *State Historical Society of Wisconsin, WHi (X3) 52034*

or mules on farms where the Cub was the only tractor. In 1945, 30 percent of the American market, about 1.6 million farms, still used draft animals. Cotton, tobacco, poultry, and vegetable-truck farmers favored the Cubs, as did people who farmed part time or maintained large gardens. The largest proportion of buyers had farms of 10 to 19 acres (4 to 7.7 ha).

At the other end of the size spectrum, McCaffrey found a growth industry in construction. To reward his work, IHC's board elected McCaffrey president and chief operating officer in 1946. Yet construction equipment was a tough market. It was dominated by Caterpillar, which sold $230 million in 1945, compared to IHC's $35 million. Cat had a postwar advantage from GI heavy-equipment operators who told their peacetime bosses what to order.

Having previously sold trucks in Ohio before moving to IHC's Chicago headquarters, McCaffrey had a feel for construction. As with trucks, bigger was better and more powerful was more useful. IHC's attack on Cat came through a $30 million investment that included acquiring a former war plant in Melrose Park, Illinois. In 1947 IHC's Industrial Power Department introduced the TD-24. Conceived by Sperry and designed by Baker, the 36,000-pound (16,329 kg) giant developed 148 drawbar horsepower. It provided buyers with 10 horsepower more than Caterpillar's D-8. The Sales Department was pleased.

Fowler McCormick had other ideas and was influenced by General Motors' brilliant chairman, Alfred P. Sloan. Sloan's mission for GM was to provide "a car for every purse and pocketbook." He inspired generations of businessmen. IHC was the "General Motors" of farm equipment, producing tractors for every farm and function.

In the mid-1940s, taking a cue from GM's separate divisions, McCormick strengthened the Farm Tractor, Industrial Equipment (encompassing portable power units, industrial, and construction machinery), Motor Truck, and Refrigeration Divisions. IHC also had a Steel Division, and Fiber and Twine (which supplied his harvesters). He gave each its own

farmalls with style

tHE CONTINUING EFFECTS of the Great Depression caused the manufacturers of durable goods to consider all means to spark the sales of their products. Some companies found a positive response to sales through styling (a method of making a product's design more fashionable and its form more functional) and through clear differentiation of a product from its competitors. Such changes at this time reflected optimism about the economic conditions in the late 1930s.

Tractor manufacturers cautiously approached styling. The crusty and ever practical farmer tended to resist frills if he thought them unnecessary, if they increased the price, or if they made the tractor harder to service. IHC engineers, working on their next generation of tractors, retained the services of famous industrial designer Raymond Loewy as a styling consultant.

The Model A was the first of the Letter Series Farmalls from International Harvester that came out in 1939 with styling. The Farmall A was also the first "offset" tractor, a Loewy concept, with the engine and drivetrain offset to the left, with the driver and steering wheel offset to the right, thus giving the driver an unobstructed view of the cultivator shovels. Close cultivation of sensitive crops was still an important job in the late 1930s before the advent of chemical weed. The Farmall B was mechanically the same, but the A had a wide front end with one long rear axle on the right and one short one on the left. The B had a tricycle front and used two long rear axles. The driver position was the same on both. As with the other Farmalls, the upgraded version of 1952 was called the Super A. It had optional hydraulics and increased engine power. The B had been dropped, replaced by the Farmall C. — *Robert N. Pripps*

left: The 1948 Super AV. The Super A appeared in 1947 and remained in production into 1954. These models offered an electric starter and lights, and they introduced the production version of the Touch Control hydraulic system.

Critical viewers will note that the word "Super" is missing from the nose decals. While the serial number 255558 confirms it is a Super AV, a mix-up at the restoration shop set the wrong decal in place.

right: The Touch Control hydraulic system consumed the limited additional horsepower that Super As had to offer over the Model A. But few operators complained once they got accustomed to the accurate depth and lift control the system provided.

experimental and research departments, sales organizations, personnel, and administrative offices. This system was extraordinarily costly because it duplicated many office functions.

Sloan's interpretation of decentralization left one central leader to make course corrections. Chairman McCormick, by several accounts, created a different system. Sloan was a director who trusted his vice presidents and managers to make everything but the most critical decisions or long-range plans. For these he was always and immediately available. McCormick's variation put more trust in his subordinates because the chairman often was unreachable and far away.

Overworked during war years, McCormick spent months in Switzerland with Dr. Jung, the psychologist and psychiatrist who

was his mother's favorite. Then in 1947 he developed pneumonia and nearly died. For reasons never explained, he kept this secret from his executives and directors. Relocating to Phoenix, Arizona, he was technically an active board chairman. He communicated with McCaffrey by phone or mail. Yet, without instantaneous communications from McCormick, McCaffrey followed his own inclinations more easily.

McCormick's absence confused his directors. They concluded he didn't care about IHC. When projects drifted off course, few people alerted him. McCaffrey's job grew as he handled problems such as retooling Louisville Works for peacetime work; labor and facilities costs tapped budgets of raw materials for tractor manufacture. A troubling sign was how IHC's share of the tractor market had slipped from 39 percent in 1940 to only 31 percent in 1949 even as the overall market grew.

The figures were grim; reality was getting worse. Delays in tooling up Cub and Model C production threw off outside parts suppliers. Searching for revenues, they bid other projects that started up more predictably. When IHC was ready to assemble the Louisville tractors, suppliers often couldn't comply. IHC's huge labor force, supported by union contracts that limited the corporation's flexibility, was not always available when parts arrived.

One other labor problem confronted the Tractor Division, at this point run by Hale, another Sales vice president without manufacturing experience. Between 1950 and 1954, Farmall and Louisville plants suffered a high annual employee-turnover rate due to resignations or layoffs from production slowdowns or assembly-line changeovers. From the beginning of each year to the next, two out of three employees changed.

McCormick blamed McCaffrey for the company's failings. But McCaffrey had an almost insatiable hunger for power. By early 1951, McCormick tried to force a showdown, but the board, especially the older outside directors, resented his intrusion. They chose McCaffrey, granting him McCormick's chief executive officer title as well. McCormick was forced into an inactive EC and board position. Meanwhile, the board promoted Brooks McCormick. Previously, he was joint managing director of British operations at Doncaster.

John Wagner's beautifully restored diesel 1949 Model MD. IHC built the first one of these on January 13, 1941, and continued to produce them until late March 1952. The D-248 inline four-cylinder used 3.875-inch (9.8 cm) bore and 5.25-inch (13.3 cm) stroke. At 1,450 rpm, the engine developed 27.5 horsepower on the drawbar and a peak of 38.2 PTO horsepower.

HALE'S TRACTOR DIVISION'S highest labor turnover was at Louisville—72.4 percent annually. Production and labor conditions there led McCaffrey and the EC in 1953 to consider moving the entire production line of the new Farmall 300 to Kentucky from Farmall Works. A study indicated that such a move would save IHC $400,000 a year in tractor shipping costs, but the relocation itself would cost $5 million.

McCaffrey's enthusiasm for construction equipment fueled expansion and improvements in IHC's crawler line. Impetus for hydraulic transmission development came from the field. The TD-24's transmission was created in cooperation with the Allison Division of General Motors. Throughout late 1947, the big crawler's insufficient development time began to show. Gears overheated, failed, and, in some instances, shattered inside the cases. Increasing financial constraints meant the Engineering Department built fewer prototypes and had shorter testing periods than necessary on new projects.

Gross sales from the Tractor Division for 1953 reached $257.6 million, but that dropped by $100 million in 1954 to $156.6 million. Shipments to dealers fell by almost half, from 14,601 in 1953 to 7,952 in 1954, taking division net income down steeply from $18.9 million to $5.4 million. Truck sales beat farm equipment for the first time in 1954, and would exceed it by half again in 1956. Budget tightening resulted from price cuts on Farmall Hs and Louisville tractors because IHC began slicing inventories before releasing the 100-Series.

The McCormicks had a noblesse oblige approach that gave the message to their directors that "we have others to do this work with us. We trust their abilities as much as our own." This behavior started with Cyrus Sr., who had his younger brothers, Leander and William, run things. Cyrus Jr. had Funk and then Legge. For Fowler it was McCaffrey. They behaved as though programmed to give others responsibilities, to seek diversity in management, often letting others take the credit. This "shared responsibility" style of directorship built the company but ultimately hastened its end.

IH
McCORMICK
FARMALL
M
FARM

Electric lighting was optional on the $3,145 base MD. The five-forward gear transmission provided a transport speed of 16.375 mph (26.4 kph).

opposite, clockwise from top left: **The 1949 Farmall Cub on Stilts.** The Tractor Stilts Company of Omaha, Nebraska, produced its first ultra-high-clearance (60 inches or 152.4 cm) conversion in 1948. Soon after, it began to manufacture kits for nearly every tractor make and model. Farmers used these conversions frequently for detasseling corn.

The 1952 Super M. IHC's concept of improving power continued with the Super M, an increased-output version of its most powerful workhorse. With 42 drawbar horsepower, this new machine was nearly 30 percent more potent than its predecessor.

The M and Super M weighed 5,100 pounds (2,313 kg) with both 6-volt electric starting and lighting systems. The tractor stood 79 inches (200.7 cm) tall and nearly 135 inches (342.9 cm) long.

The M-diesel (MD) started on gasoline and then, once its cylinders reached operating temperatures, the engine ran on diesel fuel. IHC rated both the Super M and the Super MD with identical drawbar horsepower (42) and belt-pulley or PTO (47.5) output.

The 1952 Super M. Four cylinders of 4-inch (10.2 cm) bore and 5.25-inch (13.3 cm) stroke yielded a 264-cubic-inch (4,326.2 cubic cm) engine. Running at 1,450 rpm, the engine would develop 47.5 horsepower on the PTO shaft.

In mid-1954, as introduction of the new Farmalls approached, IHC began to feel the cash crunch. The company responded by shipping tractors rapidly and randomly. In August, McCaffrey received letters from several branches complaining that "a sizable quantity of tractors [had] been shipped to territories for which they were not suited." He learned from General Sales that "errors had been made in this distribution but . . . many of the tractors originally shipped had been sold and [they were] confident that the remaining units would be moved in due time."

By October 1, 1954, McCaffrey recognized that IHC's sales projections of $1 billion for the year were too optimistic. Peter V. Moulder, executive vice president for Tractor and Implements Divisions, reported that even though the Louisville plant could produce more tractors, there were no customers. In 1953, 42

MD
MCCORMICK
FARMALL

making a small tractor larger

The 1950 Model C Demonstrator. IHC assembled these tractors at its new Louisville, Kentucky, plant. For three months in 1950, the factory manufactured tractors in white. Dealers could order them complete with cardboard placards that showed off every new feature.

IHC introduced the Model C in 1948. Yet a corporate promotion in 1950 launched a series of white demonstrators to explain to farmers that this C was something special. Touch Control hydraulics first appeared on the C- and Super A-Series tractors.

THE FARMALL MODEL C, introduced in 1948, was a replacement for the diminutive Model B. The C featured an operator platform, providing a flat floor for the control station. With the new platform, the seat was now on-center, giving the operator a view along both sides of the engine. The C was noted for its high steering wheel angle. The steering shaft ran along the left side of the engine through two universal joints. The axle arrangement of the C was like the old F-12 and F-14, which used straight axles without end drop gearboxes. This axle arrangement allowed for adjustability of the rear wheel treads through sliding hubs. What made the C look so much bigger than the B it replaced was mainly due to the larger rear tires and also because the operator sat up so high.

The Farmall C retained the same 113 ci engine as the B but governed at 1650 rpm (rather than 1400). The same four-speed transmission of the B was used. The C could be equipped with Harvester's Touch Control hydraulic system, which used an engine-driven pump.

The Farmall Super C replaced the original C in 1951. It featured disk brakes and a 123 ci engine still operating at 1650 rpm. The new engine gave about a 15 percent power increase. Hydraulics, starter, and lights were now standard, as was a new spring operator seat with a double-action shock absorber. In 1954, the Farmall Fast-Hitch made its debut on the Super C.

Production of the Super C ended in the United States in late 1954 but was then produced in France through 1958 as the Farmall FCD (French-Model C-Diesel). The French-built diesel displaced 123 ci, the same as its American counterpart.

— *Robert N. Pripps*

percent of IHC's farm tractor sales were in Farmall A, C, and Cub ranges of 9- to 24-horsepower tractors; 30 percent went to Super M sales, the 40-horsepower-and-up-class tractor. The remaining 28 percent was split between H (25- to 29-horsepower) and Super H (30- to 34-horsepower) models. The new Farmall 300, replacing the Super H, would move a tractor into the previously vacant 35- to 39-horsepower bracket.

Even as IHC broadened its product lineup, it missed an important trend. Industrial growth during and after World War II pulled 1.5 million families off farms. Many who remained bought their neighbors' land and needed bigger equipment to work holdings that encompassed a half section or more. By 1954, some 130,000 farmers or ranchers worked 1,000 acres (405 ha) or more. Owners of fewer than 3 percent of the farms bought 9 percent of the tractors in the United States.

Taking cues from the auto industry, IHC introduced a new, improved tractor model every other year. As engine

developments increased tractor performance, some horsepower categories filled and others emptied. In several instances, farmer demands for tractors sent IHC scrambling, as happened in mid-1955. Certain territories sold high volumes of diesel M and M-TA (Torque Amplifier) tractors, but IHC ended production of those in October 1954. While Engineering developed the Farmall 350 Diesel (using a direct-start Continental engine) for 1956, McCaffrey authorized the Tractor Division to acquire outside-built engines. R. M. Sheppard, Cummins, and Detroit Diesel eventually provided repowered engines for larger 450-Series models.

McCaffrey's model proliferation blurred lines between McCormick-Deering-Farmall tractor lines and the International Utility models. IHC was aware that Utility buyers used the solid-axle tractors for farming purposes. This continued with 300 Utility models, the W-4 replacements, to such extent that 87 percent of first-year sales went onto farms. R. W. Dibble, general

IH
M-TA
FARMALL
WAGNER
-1939-
ARMSTRONG

sales manager, likening these to Ford-Ferguson N-Series tractors, encouraged dealer strategies to further promote it. The Tractor Division added diesel-engine and high-clearance versions to the International 350 lineup. These duplications began to resemble the overlap of Titans and Moguls.

Steadily decreasing revenues strained budgets. Debt increased from building and equipping factories and acquir-ing outside resources. These, coupled with a sales-force-driven corporate strategy, forced management to tighten development schedules further. By the late 1950s, this issue arose regularly in EC meetings. Referring to cylinder head difficulties with diesel engines used in tractors, combines, trucks, and power units, R.

M. Buzard from National Sales challenged McCaffrey over "the possible impairment of new product development as a result of the demands made on the time of Company personnel to assist in the correction of current problems. Such practices serve to extend current product difficulties to the future and the realiza-tion of future sales [is] dependent upon the early introduction of new equipment."

This was the conundrum facing McCaffrey: Diminishing resources forced him to cut corners. His sales background created sympathy for IHC personnel who moved product. He had little understanding of those who invented or manufactured it. To McCaffrey, when a prototype or two worked, especially

left: From the flywheel forward, the Super M-TA was a pretty simple machine. The 264-cubic-inch (4,326.2 cubic cm) inline four-cylinder engine was no different from non-TA-equipped M or Super M models.

right: The 1954 Super M-TA. David Bradford of Warren, Indiana, kept an eye on the furrow as he plowed with a Model 70 3-14 plow. Engaging the Torque Amplifier reduced ground speed by about 32 per-cent while increasing pulling power by nearly 48 percent.

when they incorporated proven technology, there was no reason to delay production. Sales, to justify accelerated development requests, began exaggerating sales potential. Manufacturing designed the assembly line and ordered raw materials to meet sales projections and it "priced" the tractor (or truck, combine, crawler, or refrigerator) accordingly. Labor needs were organized prior to assembly.

In February 1956, Mercer Lee, from Finance, gave McCaffrey some sobering figures. "The present production program is in excess of the revised sales estimates by 17,000 trucks and exceeds current retail sales experience by 15,000 tractors," he said, and "current material stocks were estimated at $26,000,000 in excess of the budget." From a financial perspective, IHC was nearly out of control. Eventually, sales reports would come in two forms: projected and estimated. The former was the number hoped for; the latter was the realistic expectation. Surely somewhere they had a column labeled "actual."

IHC's sales slipped where it made its largest investments. Refrigeration products never reached mainstream urban retail outlets such as Sears, Wards, and Penney's. Company stores were in the country; once farmers had refrigerators and freezers, that

The Super W-6-TA Diesel's hefty 264-cubic-inch (4,326.2 cubic cm) engine developed 43.8 horsepower at the drawbar and 48.5 off the PTO. With five speeds forward and the TA doubling that potential, there were few conditions that would trouble this tractor.

market (barely 2 percent of the nation) was saturated. McCaffrey faced becoming a "white goods" maker, offering stoves, sinks, washing machines, and dryers—another huge capital investment. A merger of big manufacturers into a conglomerate owned by Sears, Roebuck & Co. (much as IHC was in 1902) claimed most of the business. McCaffrey unloaded the struggling Refrigeration Division for $19 million in 1955.

INTERNATIONAL
140

SEVEN

1955–1965

ENGINEERING RENAISSANCE

A four-speed transmission gave this compact tractor working range from 1.9 mph (3 kph) up to 12.8 (20.6). Its offset seating position first appeared on the small Model A and B tractors more than twenty years before this.

McCaffrey loved construction equipment. Yet agricultural implements, one of IHC's two "core" industries, remained a mystery. Once he authorized manufacture, he reassigned the same engineers to work on new products. If customer products failed, the same engineers still had the new projects while needing to create fixes as well.

Farm Equipment Division saw the need for big horsepower machinery and worked with IHC's subsidiary, Frank G. Hough Company, to create this four-wheel-drive prototype in 1959. Hough engineers gave it two- and four-wheel steering and the capability to crab. But it needed much more horsepower. *State Historical Society of Wisconsin, WHiM90-048/430/96*

By the late 1950s, due to rapidly expanded product lines and burgeoning problems, IHC was tilting off balance. On May 4, 1951, the board of directors had elected McCaffrey its chairman and chief executive officer, presenting its ultimate rebuke to Fowler McCormick. Moulder became IHC's president. In fiscal 1954, gross sales fell from $254 million to $166 million. Net income plummeted from $10 million to $2.5 million. Estimates for 1955 hit only $1.7 million. Every division cut expenses for labor, inventory, and experimentation. Despite this austerity, nothing appeared promising to the finance people.

In April 1956, the board renamed the Industrial Power Division the Construction Equipment Division (CED). McCaffrey's cherished TD-24, the flagship of his fleet, sold 1,136 in 1953 (at 90 percent factory capacity) and produced $1.8 million gross income. For 1955 it reached only 60 percent capacity and 1957 looked no better. The only way to achieve real savings was to cut product lines. McCaffrey could not do this yet.

Engineering let outside manufacturers do development. Johnston, by now vice president of Engineering, worked with Frank G. Hough Company of Libertyville, Illinois, on a prototype four-wheel-drive (4WD) farm tractor in late 1956.

Shortened Advanced Engineering cycles wreaked havoc on IHC's reputation: The large turbocharged-diesel 817 engine,

McCaffrey acknowledged the need for a balance between experimentation and development in Advanced Engineering and Product Engineering. "When a decision has to be made on a Product Engineering program," he wrote to the Farm Equipment Division (FED), "it should be expected that enough work will have already been done by Advanced Engineering to provide a sound basis for judging the feasibility of proposed Product Engineering work." For 1956, the ratio was $600,000 for Advanced and $2.5 million for Product. Then he canceled all the work of Advanced Engineering for 1957.

McCORMICK
IH
NO 120A COTTON PICKER
McCORMICK
120-A
COTTON PICKER

left: Hydra Touch was the latest version of the Touch Control hydraulic system first developed on Frame-All prototypes. By 1958, this was a very sophisticated system.

right: The 1957 Farmall 130 High Clearance. IHC introduced in 1956 the Farmall 130, the 130 High Clearance, and the International 130. They remained in production into 1958. IHC manufactured just 1,057 of these compact high-clearance models. Ohio collector Jay Peper found this uncommon original and added it to his fleet.

in development for five years, still scuffed pistons. The CED's ongoing request for a smaller crawler went nowhere until Tractor Engineering proposed modifications to its new Model 340.

Farm tractors fared no better. The EC cut daily production at Farmall and Louisville by 66 percent over two years. By mid-1957, regular staff reductions had hammered employee morale. Build quality deteriorated. The dearth of future Product Engineering disheartened IHC's engineers, and it enabled competitors in second or third place behind them to tighten the gap.

First, Deere & Company's board named William Hewitt its chairman in 1955. Soon after he assumed his new job, a new product announcement from IHC's Farm Equipment Division crossed his desk. The motto across the bottom woke Hewitt up. In red print it said "Not Content to Be Runner-Up." He wondered why Deere had been content chasing IHC.

The second event was the rapid failure of Farmall 560 tractors in 1958. The final drives couldn't handle the horsepower and torque of their new 60-horsepower, six-cylinder gas and diesel engines shoehorned into what was basically a 34-horsepower Model M tractor.

By the end of 1958, the year IHC introduced 40-Series and 60-Series tractors in every power range farmers might need, the company had sold $391 million in farm equipment. Deere had sold $464 million. IHC was now runner-up.

IHC's board elected Frank Jenks, a former accountant, as company president in 1957 and then to board chairmanship in May 1958, when McCaffrey retired. One of Jenks's initial acts was to fire the chief engineer of the 460 and 560 project. Those deserving discipline were decision makers who slashed testing budgets and hurried product launches.

After several hundred hours, Farmall 460, 560, and International 660 tractors' final-drive gears began failing. In early 1959, nearly 4 percent of the big 60-Series sat outside dealer service doors. Bull gear and pinion sets showed "the tendency toward galling," defined as the "tearing apart of metals due to overexposure to extremely high temperatures as can occur in inadequately lubricated [systems]." Farm Tractor Engineering Department (FTED) revised bull gears, pinions, and brake shafts for every tractor that was still in production. Later it expanded revisions to include the differential bevel gears and the tapered bearings and redesigned the entire differential case.

Tight finances hampered emergency response. Engineering reported it could have new gears manufactured for the 460s by late September 1959, and for 560 and 660 models a month later. Until then, it had no replacements either for tractors in production or for the 3,000 IHC had already sold. It had to replace failed sets with identical sets it knew would fail again.

To restore farmers' faith, the EC doubled the warranty on

another last look
at original loewy styling

The 1957 Farmall 450 Demonstrator. The 450 used IHC's C-281 four-cylinder with 4.125 inches (10.5 cm) of bore and 5.25 inches (13.3 cm) of stroke. These beefy engines developed 51.3 drawbar horsepower and 55.3 horsepower on the PTO during their tests at the University of Nebraska.

The 450s stretched 147 inches (373.4 cm) long as a wide front (or 4 inches [10.2 cm] shorter for the row-crop version) and stood 80 inches (203.2 cm) at the steering wheel (plus another 12 inches [30.5 cm] to the top of the nose sign). They weighed 5,600 pounds (2,540.1 kg) dry.

tHE 450 RECEIVED the same cosmetic changes as the rest of the Farmall line for 1956, but the original Raymond Loewy styling still clearly shows through. Other changes from the Farmall 400 included a displacement increase from 264 ci to 281 ci for the gasoline, LPG, and diesel versions and the 21-gallon (80 L) fuel tank became standard on the gas and diesel versions. An 18-gallon (68 L) tank had been standard for the diesel and gasoline 400s as a cost-saving measure. Farmers complained, however, when they couldn't work all day on a tank of fuel like they could before with their Super Ms.

The 450 Diesel also saw the last of the IHC "switch-over" diesel start system in which the engine head included lower-compression chambers with spark plugs. These were fed gasoline from a small starting tank through a carburetor. A small 12-volt starter fired up the engine on gasoline, and when warmed up, a lever was thrown, isolating these chambers, shutting off the gas and spark. This also raised the compression ratio and engaged the diesel injectors. At that point, the engine continued to run as a diesel.

The 450, now in the 50-horsepower class, was available in wide or narrow front ends, and in high-clearance versions.

The British-built B-450 differed from the American version in that its diesel used a glow-plug starting system wherein electrically heated "plugs" in the cylinders helped ignite the diesel fuel for starting. The B-450 also used a Ferguson-type three-point hitch, rather than the IHC two-point Fast-Hitch.

The 450 was sold between 1956 and 1958. — *Robert N. Pripps*

those lines out to twelve months and 1,500 hours on differential and rear-end assemblies. Then, in a report meant to remain confidential, it authorized full-replacement costs and the nineteen-hour labor charge for the 594 affected 460 models and 428 of the 560 models. When the project was completed, IHC had spent more than $113,000 on 460 and 560 customer tractors alone. But the report leaked out, revealing that "the marginal status of the final-drive components on the Farmall 300-, 350-, 400-, and 450-Series tractors was also becoming apparent in the field after one, two, and three years of service." After this, no recall could save the tractors' reputations.

Almost two years into the production run of 560 and 660 diesels, crankshafts began breaking (forty-two had failed by July 1960). The much higher compression of Increased Horsepower diesels made weaknesses more apparent; this did not occur in gas engines.

But it hadn't ended yet. Customers returned to dealers for the same repair a second and third time. The payback for 1957's unbudgeted Advanced Engineering program cost a fortune. In mid-July 1960, IHC interrupted I-660 production for two weeks to revise all tractors still there. Then, in late October, IHC called

back all unsold I-660 models to the factory for disassembly. For each customer tractor, they replaced nineteen mandatory and ten as-inspection-indicated pieces at no cost. This process required seventy hours of labor. It cost IHC nearly $376,000, not including parts for 1,829 chassis and 1,667 diesel engines. To provide a diesel for small-tractor customers, the board chose to import the Doncaster-built B-275 tractor. Already available to Canadians, the Standard McCormick International Diesel headed to the United States on May 4, 1959.

On February 4, 1960, the FED approved "Federal Yellow, No. 483-21 or No. 483-23 oven-dry or No. 483-22 air-dry" paint as

The 1957 Farmall Model 400. These 400-Series tractors succeeded the Farmall M and International 6-Series models using the same C-264 engine. The 400s provided the independent PTO and the Torque Amplifier that first appeared on the Super Ms three years earlier.

The 1961 McCormick International B-275 Diesel. This utility model plugged a hole in IHC's tractor line in the United States. With 30.9 drawbar horsepower and a three-point hitch with mechanical weight transfer, live hydraulics, and an independent PTO, the Doncaster import became quite popular in the "Colonies."

standard equipment on all International 340 and 460 Industrial tractors and optional on International 240, 340, and 460 Utility models. IHC concluded it was more visible at night and admitted "yellow coloring appears to create the illusion of a more massive appearance."

On March 31, the division extended that decision to include the Cub (optional), Cub Lo-Boy (standard), International 140 (standard), and International 460-, 560-, and 660-Series Industrials (all optional). On standard yellow tractors, "The current red and white color combination will be available optionally when so ordered." (This set of rulings would create havoc among collectors and restorers for decades to follow.)

UNDER JENKS, the EC moved ahead with the Improved, or Increased Power, 240X (35-horsepower) and 340X (45-horsepower) line of tractors scheduled for production in July 1961. Part of the improvement was to field a three-point hitch and hydraulic system equal to or better than the Ferguson system.

The EC included these on Improved 240X and 340X tractors tentatively set for July 1961, and on the 460X and 560X models for 1962. The new models, designated the 404 and 504, retained 240- and 340-Series styling until introduction in 1962 of the 706 and 806. All four lines appeared in the new bodywork of the large tractors.

Even though Jenks had tightened budgets, he still intended to improve products. IHC's oldest factories, the McCormick, Milwaukee, and Rock Falls Works, were inefficient. He closed the McCormick and Rock Falls plants, leaving his successors to deal with Milwaukee.

At the same time, Jenks enlarged the Engineering staff and demanded more complete testing. After the 60-Series disasters, Jenks pushed every new line introduction back a year, because "there would have been an insufficient length of time to do an adequate job of testing."

In mid-March 1960, International Harvester Experimental Research (IHER) began work on a full hydrostatic drive that replaced clutches, spline shafts, axles, gears, and other components for propelling conventional tractors.

tHE MODEL 240, like all IHC tractors for 1958, looked entirely new. The biggest change, besides the new square styling, was in the operator position and steering. The operator position now reflected the influence of the successful Ford and Ferguson tractor's "utility" configuration. Although the Farmall 240 retained its row-crop identity, the operator now sat forward of the rear axle, straddling the transmission and drive line, rather than on top with a proper platform. Because of the use of hydraulics to control implements, it was no longer necessary for the operator to sit to the rear in order to reach implement handles. The steering also reflected the Ford and Ferguson approach. The steering wheel shaft drove a worm (or ball) gear, which pulled or pushed a drag link connected to a pitman arm at the front axle. This produced a rather flat steering wheel that was characteristic of Ford tractors back to the original Fordson of 1917. Under the hood, the 240 got a governor setting of 2000 rpm and a 12-volt electrical system.

When the 340 came out in late 1957, it was a totally new design with the new square styling. Resembling an upscale 240, it was offered in a variety of configurations, including an International Utility model, a grove model, and the T340 crawler. In addition to the 135 ci gasoline engine, a new diesel of 166 ci and 2000 rpm was also offered.

The Farmall 340 was the first IHC tractor to offer the Ferguson-style three-point hitch.   A new hydraulic system used transmission oil as the hydraulic fluid. The hydraulic pump, either in the transmission housing or on the engine, was an option. The two-point Fast-Hitch or the three-point hitch were also options, as was the T/A powershift. — *Robert N. Pripps*

The 1958 Farmall Model 240. Vercel and Marilyn Bovee's 240 Row-Crop shines in the late fall sunlight. IHC kept these models in production from 1958 into 1961, manufacturing a total of 3,710.

The continuously variable ground speed at steady engine speed allowed PTO-driven attachments full independence. For plowing, rototilling, or snowblowing, with engine speed set for maximum torque, the hydrostatic drive made full engine power available from zero miles per hour up to maximum ground speed. IHER devised the systems in January 1959 and tractor design began in June. Engineer E. Jedrzykowski led the group that installed matching motors for each rear wheel, "used in parallel which eliminates the need for a differential." Jedrzykowski specified 188-cubic-inch (3,081 cubic cm) radial motors built in England and fed by a variable-displacement pump. Main power for the hydrostatic pump came from an 80-horsepower Solar Industries Titan T62T single-shaft gas regenerator turbine. The tractor, designated the HT-340 (Hydrostatic Turbine), first ran on December 7, 1959.

The hydrostatic drive eliminated the shock upon engaging forward drive under heavy implement load, minimizing strain on drivetrain components. After IHER concluded its development work, it repainted the prototype's sleek fiberglass body created by IHC's chief industrial designer Ted Koeber. For 1962 IHER added a three-point hitch, larger tires, and rear lights; it also stabilized steering, slightly desensitized controls, and transformed the prototype into the HT-341. On September 1, 1967, after years of field tests and demonstrations, and static displays, IHC donated it to the Smithsonian Institution in Washington, D.C.

Elwood four-wheel-drive conversion kits soon came to the EC's attention. Farmers who owned later-production Farmall M, 460, and 560 tractors were converting these machines to aftermarket 4WD. The EC reinstated the Frank G. Hough Company's development programs on an IHC-produced true 4WD. New chairman Harry O. Bercher (who replaced the retiring Jenks in May 1962) asked IHER to test the Elwood and other kits. IHER found them weaker in durability and performance than a comparable-horsepower, true 4WD.

Throughout this period, small-farm operators continued to sell out to larger neighbors. On February 1, 1962, the EC launched a two-tractor program with 105-drawbar-horsepower

In IHC's plants, workers busy making half-tracks and torpedoes for the war effort found it difficult to remember farming as usual.

diesel engines driving through sliding gear transmissions, to be designated the International 4100 Four-Wheel Drive tractor. Co-developed with its Frank G. Hough subsidiary, the prototypes began testing in August 1962.

Late that month, FED authorized high-clearance versions of the new 504 Farmall following requests from cane farmers in Mississippi and vegetable farmers in Florida. In October, after introduction of the 404 models, dealers telegraphed to IHC their disappointment "that the tractor did not have a constant running, or independent, type of PTO." Cost analysts determined that IHC needed to sell 636 of the 404s with live PTO to repay development and tooling costs; Sales estimated the company would sell 1,500 tractors with live PTO. Bercher, who shared Jenks's support for and belief in IHC's complete revitalization, approved it. But this was barely six months into Bercher's job, late in October 1962. The Farm Equipment Research and Engineering Center (FEREC) was at work on the Hough-based 4100, 4WD prototype.

FED began planning the next large two-wheel-drive (2WD) tractors, diesel engine only, producing 120 PTO horsepower with 12,000 pounds (5,443 kg) of drawbar pull. This four-wheel, non-Farmall-type tractor had a three-point hitch, Torque Amplifier from the 806, hydraulically actuated 1,000-rpm PTO, and a dual-system two-way hydraulic drawbar to push implements down as well as lift them. Engineering began design layout on January 2, 1963. The EC allocated $800,000 for design,

prototype assembly, and testing. FED had three prototypes ready in September. One prototype went to the Engineering Center for 2,000-hour-endurance track testing where it showed 98.2 drawbar horsepower while the other two went to Pecos, Texas, for customers' use. In Texas, one 4100 did heavy-duty deep plowing; the other did ripper and land-leveling work. They were good, powerful tractors—almost too powerful, in fact.

Tire technology held up the 1206. When the turbocharger spun the diesel engine to full power, tire sidewalls buckled; prototypes peeled lugs off treads and wheels spun on the tire beads. Project engineers from Firestone and Goodyear redesigned the tires to grip as well as unload mud. They developed an 18.4x38 heavy-duty tire specifically for the 1206. Once the company had tires that would stay underneath their tractors, production on both a Farmall and an International 1206 Turbo began in July 1965.

Early in 1964, the EC agreed with FED that there would be benefits in standardizing tractor designs for all of its markets. These two groups birthed the concept of World Wheel Tractors,

The 1959 International 660 Wheatland Diesel. This 9,875-pound (4480 kg) machine worked effortlessly through dry soil in Central Indiana. With 64.4 horsepower on the drawbar, the 36.8 drawbar horsepower high-clearance tractor provided ten forward speeds because of the Torque Amplifier. Operators could run from slower than 1/6 mph (0.27 kph) up to 15.5 (25 kph) on the road in transport gear.

In a blurring of the lines, Doncaster supplied diesel 454s to Canada, while Louisville provided all other models to Canada. Gas engines came from Louisville, while IHC's Neuss Works in Germany produced diesels. FED offered tractors either with mechanical or hydrostatic transmissions. FEREC designed body-work to create a "family styling" appearance.

The trend toward larger farms needing more powerful trac-tors became clear to FED product planners when they read sur-vey responses from 615 farmers in early 1963 about their future tractor wants and needs. Fifty percent of the respondents ran farms larger than 500 acres (202 ha), and 30 percent claimed gross sales beyond $40,000 the previous year. They averaged four tractors per farm. Only 11 percent owned tractors with more than 70 horsepower in early 1963, but 29 percent felt they would need that power in five years. On farms larger than 1,000 acres (405 ha), 58 percent needed 4WD, either below 65 horsepower or with 95 horsepower or more. Gasoline was still their fuel of choice for up to 50 horsepower; between 30 and 70 horsepower, owners preferred liquefied petroleum gas (LPG), while users above 60 horsepower wanted diesels.

The survey also told IHC that owners of tractors with less than 40 horsepower were more satisfied with their machines than those with more power. "This may suggest," the uniden-tified researchers concluded, "that the demand [for] tractors is dividing into two classes: small tractors and large tractors."

and set up development programs for a 40- and 50-PTO-horsepower tractor through joint production in U.S. plants and at Doncaster, England. FEREC, at its Hinsdale facility, designed the tractors, and by the second meeting, April 15, 1965, the large tractor had increased to 52 horsepower. FED decided soon after to manufacture complete tractors for IHC's overseas markets at the Doncaster Works, while Louisville Works would complete seventeen varieties of partially completed tractors for delivery on skids. IHC introduced these as the International 454 (40-horse-power farm-utility), 2454 (industrial-utility), 574 (52-horsepower farm-utility), 2574 (industrial-utility), and 574 (row crop).

IH
INTERNATIONAL
140

new six-cylinders

The 1961 Farmall 560 Diesel Demonstrator. Wilson Gatewood's demonstrator model showed off its gold-painted four-bottom Super Chief plow. IHC introduced the 560 in 1958 and kept these nearly 60 drawbar horse-power machines in production into 1963.

ON THE FARM in the late 1950s, more and more of the small acreage spreads were being sold to neighbors who had the desire to expand. This happened as older farmers retired and found their children were not interested in doing the hard work for the small payoff. The economies of a larger scale operation made the remaining farms practical, and the new machinery available allowed the farmer to handle more acres without hired help. Because there were no more hours in the day, work had to be done at a faster rate. That translated into a requirement for more horsepower.

In the summer of 1958, International Harvester unveiled its new line of big tractors. The first to make the scene was the six-cylinder Model 560. At its rated 60 horsepower, the 560 was touted as the world's most powerful row-crop tractor.

Later, an add-on turbocharger was offered for the 560 Diesels. It was not intended to raise the horsepower, but to improve fuel consumption and maintain rated horsepower at higher altitudes.

The 560 had a smaller sibling, the 50 hp six-cylinder Farmall 460. These were some of the handsomest tractors ever produced. The all-new styling featured long, powerful hoods. Both featured internal hydraulic pumps, seats with backrests, and 12-volt electrical systems. Gasoline, LPG, and diesel versions were available for both. For its diesel engines from 1958 on, IHC abandoned its switchover starting system. The new diesels relied on glow plugs. The five-speed transmission with Torque Amplifier powershift auxiliary was standard.

The 460 and 560 were built from 1958 to 1963. Their early reputations were tarnished by drive line failures, which were later corrected. — *Robert N. Pripps*

Bercher invested in his business, pouring money like molten ore into Wisconsin Steel, IHC's boutique mill, and even more into the Construction Equipment Division. He moved the company from its longtime world headquarters at 180 N. Michigan Avenue north across the Chicago River into the brand-new Equitable Building at 401 N. Michigan, on the site of Cyrus McCormick's first reaper factory.

With more successes than failures like 560s and 660s, the customers and the stock market remained loyal. However, IHC's resources, grown fat in the wealthy mid-1960s, were not limitless. It soon found itself laying too thin a financial blanket over its core industries.

To give small tractor sales a shot in the arm, Bercher agreed to restyle Farmall and International 140 models and the Farmall Cub and International Cub Lo-Boy in January 1963, to match 706 and 806 tractors. FED proposed an International 606 model to replace its long-troubled 460, ending production in April 1963. Manufacture began in March 1964.

Responding to increasing requests, and rejecting the Elwood system, FED produced its own 4WD tractor using a front axle developed by American Coleman Company with a differential from current production at Fort Wayne, Indiana, Truck Works. Because of its size, Farmall Works completed assembly on 706 and 806 4WD models in its Special Feature Department, established to accommodate small-series production requests. IHC offered the 4WD option with the start of regular production in August 1963.

Two years after approving development of the 120-horsepower 4WD, to be known as the Model 4300, FED shelved plans for the smaller 4100 4WD co-developed with Hough. While everyone agreed the tractor had potential, Deere introduced its 5010 with 109 horsepower and others had 4WDs closer in power to IHC's 4300.

To finish 1963, FED proposed blending its International 404 with British B-414 models to make the 424. It used the B-414's diesel engine with an independent PTO while absorbing the I-404's swept-back front axle and its better looks. This Anglo-American hybrid accepted the full range of accessories industrial customers wanted, taken from domestic parts bins. FED determined the 424 would be production-ready in nine months, around May 1964.

Two weeks after releasing the 656 to production, FED discontinued manufacturing the Farmall Cub on May 25, 1964, observing that the "agricultural use of this size tractor has declined." FED continued the International Cub, in yellow-and-white only.

INTERNATIONAL
4300
INTERNATIONAL
"BEFORE I GRE
"I WAS A CUB CADET!"
4300

In mid-June 1964, the IHER suggested replacing the existing gear transmission of the Cub Cadet with a hydrostatic transaxle. This would eliminate the clutch, transmission, axle, differential, and brake, and reduce tractor weight by 55 pounds (25 kg) in the bargain. They completed a prototype on June 7, 1963, and after tests concluded that "the Cub Cadet hydrostatic transaxle should be only the first of an expanding family of hydraulic components for use in vehicle transmissions or as hydraulic power sources for power steering, front-end loaders, bulldozers, etc." They followed it up with development plans for hydrostatic transmissions for 504, 606, 656, 706, 806, and 1206 Agricultural.

Economic conditions forced a hiatus on continued development of some tractor projects during late 1962 and 1963. The EC delayed some testing and canceled other projects outright. The Hough-designed International 4100 was shelved in September 1963 because of costs. Testing and development funds materialized, and by November 1964, preliminary structural and durability tests had only reinforced FED's belief in the tractor. Only a two-thousand-hour endurance test remained, and manufacture was scheduled to begin after that test, with sales beginning in August 1965.

For Engineering, once again, there seemed reason to have hope.

INTERNATIONAL
564
IH

1965–1975

SALES AND MARKETING REASSERT INFLUENCE

The 1969 International Harvester Company of Australia A-564. Just like the Doncaster-built B-450 models, this Geelong product made use of the gas-engine cylinder block that IHC used in its U.S. production models. With 48.5 horsepower on the PTO shaft, this was a potent machine.

During the 1960s, FED started asking questions. A new market research group anonymously polled recent IHC equipment buyers and registered owners of competing makes. IHC hoped to learn where it stood against the competition, conscious of problems in the past. The truth hurt.

The 1966 Farmall 1206 Turbo. This was IHC's first U.S. turbocharged diesel engine and its first to exceed 100 horsepower in a 2WD tractor platform.

Surveys expressed dissatisfaction, not with the equipment so much as with "the personality of the dealer," which FED's chief market researcher M. J. Steitz took to mean the professionalism and helpfulness to past, current, and future customers. Equipment drove some longtime customers away; recurring problems with 560s turned off some families, even if they had farmed with IHC since the early 1920s. Surveys urged Engineering to improve quality and Manufacturing to eliminate production defects, but mostly they stressed the importance to Sales of better attention to farmers' needs and situations.

Other responses showed a trend toward higher-speed plowing; their farmers preferred working at 5 mph (8 kph) with five bottoms instead of 4 mph (6.5 kph) with six bottoms. Farmers already were requesting tractors with 140 to 150 PTO horsepower. Many wanted dual rear tires, not only to get power to the ground without slippage but also to decrease soil compaction caused by single tires. Dealers told FED they were ready for the next Farmall.

FEREC fitted its first experimental hydromechanical transmission to a regular-production Farmall 806 early in 1965. By February 1966, FEREC released it for production. However, the EC subsequently canceled its plans, and IHC never produced the 806 Hydro.

On March 10, the EC released the F-656 and I-2656 models to production beginning October 1, followed almost immediately by the F-656 Hi-Clear. Production for the Increased Power, restyled 706, 806, and 1206 tractors, was pushed back to June 1967. IHC sold every 706 it manufactured and could scarcely afford to shut down the line. FED completed pilot models of the new 756, 856, and 1256 on September 1, 1967, and began production in October.

In May 1966, the CED found its engine lineup from 282 to 429 cubic inches (4,621 to 7,030 cubic cm) had few interchangeable parts and no family resemblance. While FED used engines by the hundreds or thousands, CED had runs of 250 or 300 scrapers or crawlers; engine manufacture on this limited scale was too costly.

Two engine "families" grew out of this research: a 300-Series, with two engines of 312 and 360 cubic inches (5,113

besting the 4010

bILLED AS "THE world's most powerful row-crop tractor at 90 horsepower," the 806 was also arguably "the world's best row-crop tractor" in 1963. It breathed new life into IHC's farm tractor operation, which had been on life support. It was all new from the ground up. It had sufficient testing to be trouble free (mostly) from the outset. When it appeared in 1963, its chief rival from John Deere, the 4010, had been out for two years. The 806 countered and trumped all of the 4010's advantages and, in 1964, Deere came back with the improved 4020.

The trend in the 1960s was for faster farming, faster plowing, and plowing with more bottoms. This meant more horsepower. Higher horsepower tractors had traditionally been the big Wheatlands or Standards, generally with only a drawbar eschewing hydraulics and three-point hitches. But now progressive farmers wanted more power from row-crop tractors.

The 806 (and the Deere 4010) used a central hydraulic system. The IHC version had three separate circuits, however, rather than the single circuit of the Deere system. The 806 had separate circuits for steering, brakes, and the implement lift.

The 806 came with three six-cylinder engine choices: gasoline, LPG, and diesel. The diesel was Harvester's first with direct fuel injection that allowed electric starting without glow plugs. An interesting aside regarding the engines is that the tractor was configured to accept the taller diesel engine. When the shorter gasoline/LPG engine was installed, there was too much daylight between the top of the engine and the hood side panels. This, IHC officials reasoned, was too reminiscent of the two-cylinder Deeres. Therefore, a taller valve cover was devised for the non-diesels.

The Farmall 806 was built from 1963 to 1967. — *Robert N. Pripps*

benefits of turbocharging

The 1966 Farmall 1206 Turbo. The 361-cubic-inch (5,915.7 cubic cm) six-cylinder made use of cylinders with 4.125-inch (10.5 cm) bore and 4.25-inch (10.8 cm) stroke. At Nebraska, this turbo diesel developed 112.6 horsepower off the PTO shaft.

a TURBOCHARGER IS A device that increases the pressure of the fuel-air mixture of the engine. It functions to achieve greater volumetric efficiency to make up for losses due to altitude and to garner more horsepower from a given engine displacement. The two main types are the exhaust-driven turbocharger and the mechanically driven "supercharger." Turbochargers have become commonplace on diesel engines to make use of exhaust energy that would otherwise be wasted and so improve the overall efficiency of the engine.

The turbocharged Farmall 1206 had the same displacement engine (361 ci) as the 806, but it was completely new. It was beefed up for the increased power provided by the turbocharger. A hardened crankshaft with larger oil passages was used. A larger radiator fan and air cleaner were added. The transmission and drivetrain were also strengthened.

The 1206 came out two years after the 806 and was styled somewhat differently. It used a welded tubular grille in place of the cast grille of the 806. At first, the 1206 was painted like the 806, but later got a gold-script "Turbo" decal on each side of the hood and a metal "Farmall 1206" emblem.

The standard transmission for the 1206 was an eight-speed sliding mesh type, with the Torque Amplifier power-shift optional. A single 1000 rpm PTO was included, as was a category III three-point hitch or IHC's Fast-Hitch, flat-topped fenders with lights and handholds, and a deluxe seat.

The 1206 was the last of the big Farmalls to be offered with a tricycle front.

The 1206 was built from 1965 to 1967. — *Robert N. Pripps*

and 5,900 cubic cm), and the 400-Series, with three blocks of 414, 436, and 466 cubic inches (6,784; 7,145; and 7,636 cubic cm). To reduce costs, CED minimized the number of engine blocks and maximized the number of parts common to each engine to provide the greatest benefits to cost, design, engineering, testing, and maintenance. CED configured the large 360- and all the 400-Series engines for production with and without turbochargers. This offered a power range from 94 to 231 gross horsepower.

Every tractor, from current-production Cub Cadets up the line to 1206s and 4100s (introduced in 1966), gained horsepower. FED also planned a new 1556 (with 140 PTO horsepower) and a 4256 (at 160) for production in late 1970. Engineers enlarged fuel tanks to allow longer working hours for farms and configured additional mounting positions to fit tanks onto tractors for those who sprayed chemicals during tillage operations.

A. O. Smith Corporation, an Ionia, Michigan–based company, contracted with IHC in March 1967 to manufacture fiberglass tractor cabs. Smith's cabs insulated operators against weather, dust, and noise, and they provided air conditioning, heat, and defrosting; tinted-glass opening side and rear windows; windshield wipers; turn signals; coat hooks; storage areas for a toolbox, lunchbox, and thermos; and space for a radio. The first prototype reached FED on February 1, 1968, and regular delivery began October 15 for Farmall and International 656, 756, 856, and 1256 models.

IHC also developed a protective frame that attached to the rear-axle carrier to address concerns over tractor rollover accidents. This incorporated a fiberglass canopy and seat belt for factory application to 656, 756, 856, and 1256 models beginning in December 1967. FEREC modified fenders and exhaust pipes that were compatible with the new rollover-protection system (ROPS).

left: The 1968 Farmall 756. The 56-Series introduced operator comfort with hydraulic seat suspension on a 38-inch (96.5 cm) wide operator platform. Gas-engine models used IHC's C-291 inline-six with 76.5 horsepower off the PTO shaft. The tractor weighed 9,483 pounds (4,301.4 kg) and sold new for $10,710.

right: IHC offered an optional two-post Roll-Over Protection System (ROPS) with a canopy for these tractors. Not many buyers went for that option. Hydrostatic power steering with a five-position-tilt steering wheel and a standard twelve-speed transmission (offering eight forward and four in reverse) made operating the 756 a delight.

100-horsepower barrier

from left: Well-known tractor-parts supplier Robert Off worked his 856 in hard, dry
soil. With just five plows behind him, Off
preferred plowing faster to plowing harder.
The 856 engine provided nearly 90 horsepower at the drawbar and 100 off the PTO
shaft. With its standard Torque Amplifier,
even six big plows were no challenge.

Twelve suitcases! These weigh something
like 75 pounds (34 kg) each. The additional
weight up front here went a long way
toward maintaining front wheel steering
under heavy plow load.

IN THE 1960S, the average farm tractor had 40 to 50 horsepower. Toward the end of the decade power was increasing dramatically, with the IHC 806 reaching 93. There were exceptions, of course,
to the so-called 100-horsepower barrier, such as the big steamers and
gas tractors like the 140-horsepower 1910 Joy-McVicker with a 3000
ci four-cylinder engine; with the 1965 Minneapolis-Moline G 706 at 101
hp, the genie was out of the bottle.

The Farmall 856 was the first naturally aspirated Farmall to
exceed 100 horsepower during its Nebraska test at 100.49 hp (Test
No. 970).

The 856 replaced the 806 in the Farmall lineup in 1967. It was
restyled to bring it in line with the rest of the "56" Series offerings and
beefed up to handle the increased power of the enlarged six-cylinder
diesel of 406.9 ci.

Comfort and convenience were fast becoming major selling factors in tractors. Therefore, the 856 featured an optional tilting steering wheel and a new hydraulic power seat. It tilted, moved up or down,
and slid back and forth. It was also equipped with folding armrests. An
improved factory cab with two doors came out in 1970. Without the
cab, factory-installed ROPS was available.

The 856 was available in tricycle and wide-front and in high-clearance versions. Gasoline and LPG options were not offered. The transmission provided sixteen speeds forward and four in reverse when
equipped with the optional Torque Amplifier powershift auxiliary.

The 856 Custom was a stripped-down economy version. It had
single remote hydraulics, single headlights, a smaller fuel tank, and a
smaller air filter.

The Farmall 856 was produced from 1967 to 1971. Production
amounted to about 25,000 tractors. — *Robert N. Pripps*

The 806–856, 1206–1256, and 4100 models benefited from a new Category II and III Quick-Hitch coupler FED introduced for use with the three-point hitches. It sought to make implement attachment possible without requiring the operator to leave the cab (except for PTO attachment). Compatibility with other makers' quick-couple implements was another objective.

The FED solidified plans for its World Wheel Tractor series. The first, a 40-horsepower medium-duty utility-type model for farm or industrial uses, was coded TX-19 internally and became the International 454. Plans included the TX-36 in International and Farmall 574 model designations, produced in Louisville for the United States and Canada, for introduction in 1970.

Beginning in 1967, FEREC used computers for experimentation and testing systems. This shortened development time considerably. Now the science of metallurgy interfaced with a body of field experience, and everything from compression spring design to V-belt life projection to rockshaft arm stress analysis to gear design and axle-bearing load assessment was possible in the computer.

IN FEBRUARY 1968, TWELVE prototype World Wheel Tractors—six each of 40- and 52-PTO horsepower units—had completed thousands of hours of testing.

A small problem arose with hydrostatic transmissions affecting forward and reverse control. It involved few 656 models. FED quickly responded with a small spring to secure the pump-servo cylinder-control valve, requiring an hour's installation

at company expense. Harry Bercher told FED that supporting Engineering was "in the company's best interest."

In May 1968, IHC elected the fourth McCormick, Brooks, as president and chief operating officer. Brooks, at fifty-one years old, was a great-grandson of William McCormick, who was Cyrus McCormick Sr.'s business manager and youngest brother.

Brooks observed that IHC derived most of its profits from farm equipment, most of its sales from trucks, and most of its expenses and losses from construction equipment. He championed the World Tractor concept as a means to pare costs by simplifying tractor lines and adopting universal parts.

The 1969 Farmall 656 with Disc. IHC offered the 656 with diesel or gasoline power. This gas-engine version developed 44.7 drawbar horsepower at 75 percent load in its tests at the University of Nebraska.The 656 gas tractors, in production from 1965 into 1972, sold new for $7,340. They weighed 6,350 pounds(2,880.3 kg).

The 1970 Farmall Model 1026 Hydro. Ron Neese's powerful 1026 Hydro waited with its Woods Batwing Model 3180 mower. The mower could cut a 15-foot (4.6 m) swath. The 1026 Hydros first appeared in 1969. In University of Nebraska tests, the 407-cubic-inch (6,670 cubic cm) turbo-diesel engines developed 110.7 horsepower at the PTO shaft. The 1026 weighed 10,400 pounds (4,717.4 kg). It sold new for $14,970.

TO SUPPLEMENT 656 HYDROSTATIC DRIVE TRACTORS, in mid-June 1968 FED proposed new 826 and 1026 models at 84 and 112 PTO horsepower for introduction November 1969. In July it released the 4156 with 140 PTO and 125 drawbar horsepower to replace the 4100 4WD model. To ensure no future problems with this power/workload increase, FEREC upgraded rear-end gears to specifications that worked with the revised 460 and 560 models.

Hydraulics technology had not kept pace with drawbar power. Tractors now could pull more than they could lift. This imbalance had less impact on fieldwork, but it placed risks on farmers driving tractors with raised implements on the roads. FEREC introduced IHC's latest weight-transfer hitch in November 1968. This system advanced Harry Ferguson's geometric A-frame structure; it transferred loads off the rear of the tractor to the steering and drive wheels. Now the hydraulics could raise heavier implements without lifting the front of the tractor. FED offered production units for tractors up to the 1256 models beginning in December 1969.

FED did not upgrade the 1256 itself because it had planned a replacement to keep up with the newest horsepower war in farm equipment. Deere's new Model 4520, with 120 PTO horsepower, forced FED to raise the ante with a 125-PTO-horsepower model, the 1456, and a later 155-horsepower model. To accommodate the additional power in the 1456, FEREC increased virtually every dimension and strengthened each element of the tractor. Engineers widened gear faces from the transmission to the rear end. They enlarged the radiator, moved it forward, fit a bigger fan, and gave it greater clearance. They increased brake disc diameter from 8 inches (20.3 cm) to 11.375 inches (28.9 cm), and enhanced every brake component accordingly. They enlarged rear-axle diameter from 3.25 to 3.5 inches (8.3 to 8.9 cm).

On February 28, Neuss Works in West Germany informed the EC that Europe needed larger tractors. Under agreement with FED, Neuss Works imported D-310 engines, which it linked to partially synchronized transmissions to produce a prototype TX-47, 80-horsepower tractor by November 1969. Regular production as the Model 846 started March 1971.

Existing tooling, factory production lines, and parts inventories challenged Brooks McCormick's dream of a World Tractor

hydrostatic farmalls

aN INTERESTING CHANGE took place in the Farmall line in 1967. The Farmall 656 was optionally equipped with a fully hydrostatic transmission. A hydrostatic transmission consists of a variable displacement hydraulic pump driven by the engine supplying a fixed or variable displacement motor driving the wheels. The hydrostatic transmission seemed to be an ideal application for farm tractors, giving infinitely variable ratios as required by simply moving the S/R (speed/ratio) lever. The two-range power-shift (T/A) was included to give full rpm speeds of from barely creeping to 21 mph (33.8 kph). On the downside, hydrostatic transmissions were only 70 to 80 percent efficient while gear transmissions are generally better than 98 percent. That translates into lower drawbar power and large quantities of heat to dissipate.

The 656 was a nominally 60 hp six-cylinder tractor.

The 544 was offered from 1968 to 1973 with either a manual five-speed transmission with Torque Amplifier or a hydrostatic transmission with a manual two-range auxiliary. Four-cylinder gasoline and diesel engines were new. The LPG option was not offered. Other features included fully hydrostatic power steering, live hydraulics and PTO, and a category II three-point hitch with draft control.

The 544 was nominally a 50-horsepower tractor.

Harvester dealers were besieged with complaints that hydrostatic-drive tractors could not keep up with gear-drive tractors of the same model number. The farmers thought they had paid a premium for the hydrostatic tractor and they should not be shown up in the field. To eliminate the problem with a minimum of cost, IHC simply rebadged the 966 and 1066 hydro Farmalls as Hydro 100s (nominally 100 hp). These tractors used a 436 ci six. The Farmall 666 was rebadged as the Hydro 70 (nominally 70 hp) without change. — *Robert N. Pripps*

IHC introduced Hydrostatic Drive on its 1967 Model 656 tractors. This system provided an infinitely variable travel speed for the tractor and implements while allowing the operator to maintain peak horsepower or torque performance through challenging crops or terrain.

The 1970 Model 1456 Gold Demonstrator. For 1970, IHC painted side panels, fenders, and hoods gold on many of its demonstrators. The 407-cubic-inch (6,670 cubic cm) turbo-charged diesel developed 131.8 horsepower at the PTO shaft.

Attention-getting paint schemes were nothing new with IHC's sales department by 1970. In 1950, the Louisville Works painted nearly its entire output of tractors and implements white for three months to promote dealer demonstrations.

line. The world had less need for machines larger than 756s, yet these and larger machines sold well in North America. Horsepower wars drove FEREC, while sales fed IHC's treasury. The company invested $24 million in tooling at the Melrose Park plant to expand annual production to 40,000 of the DT-466 diesel engines. Late in March 1969, FED authorized a two-phase replacement plan for its large tractors. With production beginning November 1, 1970, the 60-Series upgrades offered small increases in power and a new grille, hood, side panels, and overall appearance.

The World Tractor philosophy created a model for U.S. markets only. The Louisville Works product committee took requests from eastern and southeastern farmers for a 32-horsepower tractor, larger than Cub 14s and 154s, but smaller than 444 and 454 models. Doncaster provided running gear from their B-275 tractor with the front axle, grille, hood, instrument, and steering mechanism from the World TX-19 and TX-36 tractors. Louisville had a 32-PTO-horsepower tractor for regional U.S. markets as an International 354 with production starting in November 1970.

Yet the Louisville plant experienced problems with the International Cub 154 Lo-Boy, with nearly one in four units to suffer driveline failure. Tight tolerances caused clutch shafts and drive-coupler hub assemblies to wear prematurely. FED revised the Louisville assembly lines while also authorizing an industrial 25-horsepower companion to the Lo-Boy. Called the International 254, it used the hydrostatic transmission and a four-cylinder, water-cooled Renault engine. Production started in November 1972.

FED discontinued the 756 diesel from U.S. production in mid-July 1969, replacing it with the 826 using IHC's own D-358 diesel produced at Neuss, Germany. However, it continued to ship partially assembled 756 models to its plants in France, Mexico, and Australia well into the early 1970s. These plants installed Neuss Works D-310 engines from West Germany.

Big power drove product development, sometimes from outside.

IN LATE 1970, MISSISSIPPI ROAD SUPPLY (MRS) in
Flora, Mississippi, began a cooperative program with FED. MRS
developed and manufactured two 4WD articulated tractors of
130 and 155 PTO horsepower and planned to take over production of the 4156 4WD, four-wheel steering tractor for IHC.
In addition, MRS licensed its three current-production 4WD
tractors (up to 236 PTO horsepower) to IHC. MRS also gained
access to engines using CED's D-466 six-cylinder and Truck's
DV-550 V-8 for 130- and 155-horsepower 4WD models. IHC handled worldwide distribution. MRS planned production to begin
in February 1971, following a $12,000 tooling expense, to adapt
IHC engines to MRS chassis mounts. But the program collapsed.
IHC then went to Steiger in Fargo, North Dakota, after proposing
it use IHC's engines in its large 4WDs.

Big power drove product development, sometimes from
outside. FEREC engineers knew that tractor owners modified or
replaced IHC's engines to get more power. V-8 engines began
to appear not only for power and smooth operation but to give
farmers bragging rights.

In June 1970, FED authorized production on its DV-550
130-PTO (at 2,400 rpm) horsepower tractor. The V-8 appeared
in mid-October 1971 as the 1468 model, while keeping the 1466
inline-six (DT-436) in production. Two years later, a higher-speed
2,600-rpm version increased DV-550 output to 145 horsepower.

The 1971 International Farmall Model 856 with International Model 710 5-16 Plow. Out of the factory, these 856 tractors weighed 8,620 pounds (3,910 kg). Adding a little weight to the front end and slipping a dual rear tire on the land side meant that very little would interrupt the progress of this machine.

WEEKS BEFORE BECOMING IHC CHAIRMAN in May 1971, Brooks McCormick had a grand plan for simplifying product lines, but it suffered two quick hits.

The Sales Department had always rushed Engineering and Manufacturing. Jenks and Bercher together diminished Sales' influence, giving engineers time to get caught up and think ahead. During Bercher's last year, from May 1970 to 1971, even as the economy contracted, Sales resumed its old ways. If any competitor introduced something IHC did not have, it created a vacuum in IHC's line that product planners abhorred.

Describing the International 644, FED's tractor planners proposed fitting the Neuss D-239 diesel engine into a combination of International 544 and 656 components designated the International 654. "The resultant tractor," they said on May 11,

1971, "would be targeted to compete with the utility-type Ford 5000 and Massey-Ferguson 175 diesel tractors in this size and price range." Two days later, to fill a gap between the present 154 Lo-Boy and the forthcoming 354 tractor, Sales proposed importing the Kimco 242 tractor as the International 242. Kimco was Komatsu-International Manufacturing Company, a joint venture between Komatsu and IHC created in the late 1960s. Sales chased a market for 3,000 tractors to provide products for dealers who wanted to carry Japanese imports. Shortly after FED had approved these programs, Brooks McCormick took over.

He inherited a company in trouble. The corporation's entire profits, $45.2 million, just met shareholders dividend disbursements. Brooks had no reserve funds. Budgets set for 1972 and beyond had no flexibility. He inherited employee wage and benefits packages that were more generous and costly than either Deere or Caterpillar. Concerned about making IHC "well managed," he hired outside executives to bring in new ideas. Senior managers felt betrayed because they didn't get promotions.

Some issues demanded attention. The Vietnam War started an inflation that raised the consumer price index 60 percent from 1967 to 1975. Part of that surge came in February 1971 when the six-country Organization of Petroleum Exporting Countries (OPEC) agreed with twenty-three of the world's oil companies to a settlement nearly tripling fuel prices in the United States. Oil-producing countries, wealthy from sales of their natural resources, purchased U.S. produce. The Soviet Union bought $136 million in grain. Famines starved millions in Asia and Africa, which frightened the rest of the world about

ultimate farmall

LATE IN 1974 the Farmall ranks were joined by the most powerful and the most up-to-date Farmall ever, the 1566. It was powered by the DT 436 direct-injection, turbocharged 161-horsepower diesel coupled with the new 3x2 (six speed, twelve with optional Torque Amplifier) Easy-Shift transmission, driving through a newly designed drivetrain. A Category III three-point hitch and a 1000 rpm PTO were provided. A central hydraulic system supplied the muscle for the three-point hitch, the power steering, power wet-disk brakes, and the clutch.

The new component-style drivetrain allowed easy replacement of items in the final drive without disturbing others. Adjustable wide-front axle was another option available.

The 161-horsepower six-cylinder engine could not be matched by the DV-550 V-8, not leaving much of a future stablemate for the Model 1568.

The Farmall 1566 was the first Farmall to be fitted with the new Deluxe soundproofed and air-conditioned cab, complete with an eight-track stereo system. The cab was painted red with a cream roof, like the hood trim stripes.

The1566 was built from 1974 to 1976. IHC was already suffering cash flow problems and labor unrest. Some maintain that dismissing the Farmall name, which had become an integral part of the "Red Tractor" phenomenon, contributed to the downfall. True, chief competitor Deere, Inc. had dropped the "John" from their modernized name, but "Farmall" had long since lost its connection with the all-purpose machines of the 1930s. Nevertheless, Farmalls had a separate serial number system from those badged "International" until 1973. For IHC, the Farmall name disappeared completely in 1976. — *Robert N. Pripps*

hen's tooth collectible

IN 1976, THE last year for the big eight-cylinder Farmall 1568, only a few of which were badged "Farmall," marketing gurus at IHC bragged that the company had produced five million tractors since International Harvester was born on July 26, 1902. A large percentage of them were bright red and carried the Farmall name.

Loyalty is a word not so much in vogue today as it was in the twentieth century. Loyalties to one's faith, country, flag, political party, and mate were then taken for granted. Such devotion is particularly true for those who farmed with Farmalls. The reasons for this allegiance are many and hard to understand if you haven't been there. Nevertheless, the disappearance of the Farmall name was "gut-wrenching" to some and they took their allegiance elsewhere.

Fewer than one thousand of the big 1568s were built between 1974 and 1976, and only a few of them carried the Farmall badge. These are now "must-haves" for the ardent collector with deep pockets. These machines are truly impressive, especially with their red Custom factory cabs.

The 1568 was the same tractor as the 1566, except for the DV-550 V-8 engine in place of the turbocharged six. The 1568 differentiated from the 1468 models in that the engine was upgraded to just over 150 hp, a stronger planetary final drive was included to cover the increased horsepower, and a six-speed transmission replaced the eight-speed unit of the earlier series. The Torque Amplifier was a usual option, doubling the ratios available. — *Robert N. Pripps*

having adequate supplies of food. Farmers expanded their holdings and bought equipment.

There was justification for bigger, faster tractors. Within
nine months of the launch of the 4166 and initial production
of the Steiger 4WD tractors, FED had already heard questions
about maneuverability and soil compaction hounding FEREC
engineers. In response, they mated two 1066 final drives
together using a 4166 transmission and transfer case. This articulated 4WD prototype was a forerunner of the series later known
as 2+2 tractors.

The Steiger relationship proved that "badge engineering"
was convenient, and for the short run, more cost-efficient than
IHC spending millions developing their own tractors. Badge
engineering involves one company purchasing already engineered and developed products from another and then putting
its own name or badge on the item. Steiger agreed to build a
175-gross-horsepower unit (based on the twin 1066s) and a
275-horsepower version using the DT-466 engine. Steiger would
design and manufacture the transfer case, buying transmissions
from Fuller. FED designated this as the TX-111.

On April 24, 1972, FEREC introduced its synchromesh
transmission to replace the previous sliding-gear type, especially
for the 700-, 900-, 1000-, and 1400-Series tractors. This four-
speed unit provided shift-on-the-go capability under unloaded
conditions. It could not accomplish the moving shifts under full
load that the Torque Amplifier could do, nor did it provide the
variable-speed characteristics of the hydrostatic. Still, IHC was
the only maker offering synchromesh with an optional Torque

Amplifier. Production began in October 1973 for the 1566 and
1568 models.

FED released plans for the DT-466-engined 175-PTO-
horsepower 1566 and DV-550-engined 1568 models. Based
on the 1466 frame, it incorporated a modified planetary final
drive from the 1566 and 1568 models and a three-speed sliding-gear transmission. Production of this 1500-Series began in
November 1973.

Farmers noticed a change on the side of these new machines.
These and all IHC farm tractors released in late 1973 as 1974
model year production no longer bore the name Farmall. Over

Sixteen forward gears gave
the 11,860-pound (5,380 kg)
1066 plenty of flexibility, even
when towing a full corn wagon
through stubble. The tractors
sold for $21,080 new.

end of the tricycle tractor

The 1975 International 766 Turbo Diesel. IHC manufactured these tractors between 1971 and 1976. Weighing 9,538 pounds (4,326.4 kg), these diesels sold new for $15,240. The D-360 inline six-cylinder developed a robust 85.4 horsepower off the PTO. The TA system provided sixteen speeds forward and eight in reverse.

tHE REIGN OF the narrow-front (tricycle) farm tractor seemingly came to an end with the Farmall 766, the last of the breed to be tested at the University of Nebraska (Test No. 1094). The reign began with the Farmall Regular of 1924. From then until the 1970s, the true all-purpose tractor was a tricycle, although most manufacturers offered optional wide fronts. Several factors led to the demise of the "trike." First, the use of chemical weed killers virtually eliminated the cultivation of tall crops. The chemical tanks were often hung on the nose of the tractor and the increased stability of the wide front was appreciated. Also, wide fronts improved steering in soft ground and were necessary for adding hydraulic front-wheel assist. Tricycles were offered by IHC and others for several years, but the 766 was the last tested.

The "66 Series" tractors were unveiled in the fall of 1971 replacing the "56 Series." The 766, the smallest of the line, was available in gasoline or diesel versions, both producing about 83 PTO horsepower. IHC did make smaller tractors for the U.S. market, but they were in the utility configuration.

The 766, like others in the series, retained much from the models replaced. All were restyled and strengthened in the front end and frame. An adjustable wide front axle was standard, the tricycle version an option. A deluxe cab, an electrically actuated traction lock, and front-wheel assist were also options. The eight-speed sliding gear transmission was standard with the T/A as an optional add-on.

The Farmall Model 766 was produced from 1971 to 1976.

— *Robert N. Pripps*

to cab or not to cab

tRACTOR CABS TOOK a while to be generally accepted by the farmer. Before the routine use of hydraulic implements, the operator had to lean back to reach the plow, or other implement, handle, precluding an enclosure. Others felt that the cab compromised visibility, or that they missed the fresh air and sunshine. Further, cabs were supplied by other than the tractor maker; they were often inconvenient and didn't offer protection from dust, rain, and cold. Some early examples of "factory" cabs were the 1911 Pioneer and the classic MM UDLX.

By the 1970s, factory cabs were becoming a selling feature, so Harvester offered one for the 966 in 1972, the second year of production. The "Deluxe" cab had a heater. The "Custom" did not. (The Custom version was discontinued in 1973.)

International Harvester unveiled the "66" Series tractor lineup in the fall of 1971. The Farmall 826 and the 1026 were aberrations in the IHC tractor numbering system. This was corrected in 1971 by replacing the 826 with the 966 and the 1026 with the 1066. These were produced through 1975.

The 966 with its naturally aspirated 414 ci diesel was in the 90- to 100-horsepower range. In 1973, competition forced IHC to increase rated engine speed to 2600 rpm, giving it 101 horsepower. The standard transmission for the 966 was the eight-speed mechanical, with optional Torque Amplifier. The fully hydrostatic transmission with the two-speed sliding-gear auxiliary was also available.

The Farmall 1066, which replaced the 1026, used the 414 ci turbo diesel and was available with either the eight-speed mechanical transmission and T/A, or the hydrostatic drive and T/A. Optional on the 1066 was the new deluxe two-door cab with air conditioning, a Farmall first. — *Robert N. Pripps*

The 1975 International 1066. The engine for these standards was IHC's turbo-charged D-414 incline-six. This workhorse developed 125.6 horsepower on the PTO.

the previous two years, as the EC eliminated duplicate product lines, the International name grew larger on FED tractors while Farmall shrunk. At the end of 1973, it disappeared altogether.

After six months of discussing consumer products among FED product planners, they were confident that IHC's name was back in public awareness. That faith prompted new efforts to reduce the time and funding for testing new products. Sales regained much of its former stature. Product Development again assumed a catch-up role, where engineers produced IHC's version of someone else's improvement.

When FED introduced hydrostatic transmissions, it kept drawbar horsepower as close as possible to Gear Drive models so buyers could compare work potential. Engine developments increased horsepower, but the Gear Drive's greater efficiency produced higher power readings. With the 66-Series, the discrepancy grew so much that 1066 hydrostatic drawbar horsepower fell nearer to 966 Gear Drive statistics. In March 1973, FED renamed the hydrostatic tractors to obscure direct comparisons. It eliminated both 966 and 1066 models, replacing them with the Hydro 100. The 666 became the Hydro 70.

IN OCTOBER 1973, OPEC shut off its wells. That far-away act had ramifications that crashed down around Brooks McCormick's head, making tractor fuel economy a factor in new purchases.

He had many other serious distractions. The Occupational Safety and Health Administration (OSHA) was formed in December 1970. OSHA affected tractor operators in the field and engineers creating tractors for them; it also meant that all of IHC's facilities had to comply with strict regulations, no matter what it cost. The previous July, Washington had also created the Environmental Protection Agency (EPA) to give teeth to the 1963 Clean Air and Water Act. McCormick saw IHC's aging plants and its dirty, inefficient, and unprofitable foundries and steel operations as future money pits, facing cleanup and modernization costs of hundreds of millions.

IHC's profits increased in farm equipment. In 1975 sales topped $2.1 billion, and for the first time in ten years, farm equipment outsold trucks. That was a hollow distinction; the country had endured the worst of OPEC's fuel crisis through early 1974, and the Truck Division experienced a huge loss through 1975. A year later, the Construction Equipment Division lost $4.7 million. IHC's credit rating dropped to "bad risk" status. The outlook was bleak.

IN 1959, HARVESTER'S rival Massey Ferguson acquired the British Perkins engine company. Then in 1967, Massey introduced its Model 1150 tractor powered by a 511 ci Perkins V-8 diesel of 135 horsepower. Then, in 1971 rival Oliver introduced their Model 2255 tractor with a 573 ci Caterpillar V-8 diesel. This was too much for IHC's marketing gurus, who then insisted that Harvester have a V-8 tractor, too. Hence, the Farmall 1468 was built between 1971 and 1974.

The 1468 was essentially a Farmall 1466 with a DV-550 (548.7 ci) diesel V-8 engine supplied by the IHC truck division. The engine was naturally aspirated and produced a PTO horsepower of 145. To save fuel, the DV-550 was set up to run on only four cylinders at idle, or low loads. Cylinders 1, 4, 6, and 7 received fuel under these conditions, and the other cylinders had their valve lifters vented so the valves stayed closed. This economy measure was first used in General Motors Cadillac V-8 engines and subsequently in all their V-8 truck engines.

The 1468 was not available in a high-clearance version or with a differential lock, and only the 1,000 rpm PTO was offered. The eight-speed manual transmission was used along with the optional Torque Amplifier. A Deluxe cab with improved visibility and soundproofing was optional.

Just under 3,000 1468s were delivered.

It should be noted that International Harvester officially dropped the Farmall name in 1973. Some Farmall tractors in production prior to 1973 that were continued beyond 1973 carried the Farmall badge to as late as 1975.

— *Robert N. Pripps*

Wilson Gatewood's V-8 developed 145.4 horsepower off the PTO shaft. It was not easy making the engine fit. It was 550 cubic inches (9,013 cubic cm). It came from IHC's Truck Division, and Farm Equipment Division had to modify it to fit between 25-inch (63.5 cm) frame rails.

5488
INTERNATIONAL

NINE

1976–1999

A STUMBLING GIANT IS RESCUED

The 1984 Model 5488. The two-wheel-drive models sold new for $66,515. IHC priced the four-wheel-drive versions at $78,340.

The country's bicentennial put most of the United States in good cheer. In a patriotic flurry, FED released the 4568, its 300-horsepower tractor built on a Steiger chassis. IHC sold it only in 1976, renaming the big, articulated 4WD the 4586 for 1977.

IHC launched the 86-Series in November 1976, with two Hydro models, the 86 and 186, and a full range of small to large tractors, from the International 868 up to 1586, 4186, and 4386. These tractors adopted the A. O. Smith XCF-65 "pod"-type cab that moved the operator forward and made room for the fuel tank at the back. It also introduced a 284 from Kimco, the joint venture operation with Komatsu.

Along with more power, the 86-Series bought farmers the "Control Center," the new weather- and sound-insulated cab. It provided the farmer with more instrumentation than earlier tractors. The polyfoam and iso-mount insulators, thick carpet, and

left: IHC's D-466 engine powered the 4366. The inline six-cylinder turbo diesel developed 163.1 horsepower off the PTO shaft.

right: The 4366 sold for $37,400 in 1976, without dual wheels or plow. In that configuration, it weighed 18,800 pounds (8,528 kg).

opposite: The 1975 International 4366. IHC manufactured these four-wheel-drive models from 1973 into 1976. Frank Ferguson's tractor worked with the company's 12-foot (3.7 m) front blade. The machine ran dual 18.4-38s all around. Having just completed some road building around the farm, this machine was ready for the snows of winter.

86 HYDRO
INTERNATIONAL

wraparound glass isolated farmers from sounds that typically told them whether all was well with their machines.

Steiger developed an even larger chassis for 1979, the 4786, with 350 horsepower. IHC then began to replace the 54-Series and the 74-Series line in 1980 with the full-diesel line of 84-Series small- to medium-size tractors, reducing two lines into one. The 84-Series also provided torsion-bar draft control, planetary final drives, and the three-lever hydraulic hitch control. International 884s came with Torque Amplifier standard, while IHC offered it optionally on 584, 684, and 784 models.

Mitsubishi manufactured IHC's smaller tractors. These small diesels provided between 15.2 and 21 PTO horsepower as models 234, 244, and 254, manufactured in Japan with IHC-specified features.

FEREC, which had mated rear ends of two 1066 production tractors to create a prototype articulated 4WD, tried the same thing with 86-Series prototypes in a continuing effort to develop mid-power-range models. It designated these as 2+2s because they consisted essentially of one 2WD tractor plus another one. FEREC carried over the Control Center cab system. To use the cab and the 86-Series final drive, IHC designed the 2+2 so the solid front axle steered by pivoting the front half of the tractor. The long nose housed the engine-before-drive-axle configuration of 66- and 86-Series tractors. This required no engine redesign and very little drivetrain modification. It also increased tractor stability with heavy rear-mounted implements.

Before IHC introduced its 84-Series in 1980, it brought out the new 3388 (130 PTO horsepower) and 3588 (150 horsepower) 2+2 tractor in late 1978. Then, a year later, it introduced the 170-horsepower 3788 2+2, using the latest DT-466B turbo diesel.

SEVERAL PHILOSOPHIES EMERGED from world head-quarters as the bicentennial passed. In 1975 Brooks McCormick approached Booz, Allen & Hamilton (BAH), a management consultant firm. BAH proposed IHC incorporate its foreign plants within the divisional structure in the United States. European and Asian tractor operations all fell under Agricultural Equipment Group (AEG). Construction became the Payline Group (including smaller industrial machines). Truck Group and Solar Turbines International Group remained separate divisions.

As the farm equipment market improved, McCormick envisioned overtaking Deere as the farm equipment industry's number one manufacturer.

IHC watched Deere and Caterpillar produce big technologi-cal jumps in the 1960s and early 1970s. Experience and common sense told IHC engineers and board members that these two companies would need to leave these machines in production for ten years to pay for them. This gave IHC room to introduce innovations in the 1980s that might allow them to move ahead.

To ensure that plan continued, McCormick picked Archie McCardell as his successor. McCardell joined IHC in August 1977 as president under McCormick. Two years later, McCardell became board chairman and tagged Warren T. Hayford, who left the aluminum can industry in mid-1979 to join IHC.

McCardell came from Xerox, where careful five- and ten-year financial management and advance planning kept them in control of their market. McCormick knew IHC needed long-term planning and economic controls rather than short-term reaction funded by whatever resource was available instantly. Hayford brought with him ideas for cutting costs by increasing plant efficiency. Hayford's background, however, had not prepared him for the cyclical nature of farm equipment markets.

3488 HYDRO
INTERNATIONAL

AGRICULTURAL EQUIPMENT group sales continued to climb, yet the Truck and the Payline Groups lagged behind. McCardell committed $150 million to expand DT-466 engine production and nearly $200 million to another AEG project, dubbed TR-4 and TR-3A (the future 50-Series and 30-Series). To finance this, McCardell trimmed and clipped excesses and waste from everywhere within the corporation. He managed to whittle $300 million out of overhead and costs in 1978 and 1979.

Hayford took over McCardell's chores. He cut costs and introduced his can-industry work ethic, "the 8760 plan," representing the number of hours in the year. When Hayford arrived, IHC worked fourteen of the twenty-one shifts a week, which, to his mind, meant one-third of its capital resources were wasted in those seven idle shifts.

There were three seasons and a distinct buying cycle for farm equipment. There was no season for aluminum cans. According to IHC historian Barbara Marsh, the first season occurred prior to spring planting when about a third of annual sales took place. The second opportunity was fall harvest, and this constituted half of annual totals because of combine and other high-priced equipment sales. The last appeared in late December for year-end tax planning.

On May 19, 1977, the newly named Agricultural Equipment Group (AEG) approved an earlier FED development and testing request. That involved power trains for a series of larger-horsepower tractors, along with a new pressure-flow-compensating hydraulic system, as well as a combined project originally referred to as "MATH," for "Modular Axle, Transmissions, Hydraulics," that IHC had first used in CED's TD-20E crawler.

The Modular Axle, described in a 1978 engineering data report, was a "final drive on two-wheel drive tractors . . . this assembly in modular axle form will be used as a front and rear unit on large four-wheel drive articulated tractors."

The pressure-flow-compensated pump provided low hydraulic pressure for steering, transmission, a hydraulic-oil cooling system, independent PTO, and wet multiple disc brakes. The high-pressure side took care of lubrication, draft control, and auxiliary valves. AEG rated the transmission at 200 horsepower, needing only to revise PTO clutches to accommodate even more power. FED planned to produce three models. One would rate 135 PTO horsepower; the second would develop 160 horsepower; and the largest had 185 horsepower using Melrose Park's new DTI-466 turbocharged and intercooled diesel.

The greatest technical advancements were in transmissions offered with the 50-Series TR-4 and the lower-horsepower-range 30-Series, known inside AEG as the TR-3A tractors. For the 3A-Series, FED had in mind the new P-3A Constant Mesh (synchromesh) transmission. For TR-4 models, FED was completing either a Synchro-Torque or a Vari-Range transmission. Engineering planned to introduce these transmissions in the 2+2 models as well. The Synchro-Torque transmission required a clean lab "white room" to assemble the internal parts. AEG now had taken hold of the technology of the late 1970s and was looking far ahead.

The Farm Equipment Research and Engineering Center, FEREC, chose its DVT-800 V-8 turbo diesel to power these machines. With actual displacement of 798 cubic inches (13,077 cubic cm), these hard workers churned out 265 horsepower at the PTO.

By year-end, sales from all groups had reached record levels. McCardell's economics and Hayford's cost cutting had provided the corporation its highest profit margin in ten years. McCardell restructured some of IHC's debt into short-term financing, terminating at the end of 1981, rather than relying on the typical long-term arrangements used for such expenses as new factories and tooling.

Manufacturing in all groups increased production through the spring and summer in advance of labor union contract negotiations and a strike by the UAW (United Auto Workers) that IHC management expected. According to historian Barbara Marsh, IHC added "$125 million of additional inventory to sell in case the strike went on any period of time."

The UAW walked out on November 2, the day after IHC published annual figures. The 35,000 strikers received $50 per week. It hardened their resolve against IHC.

The strike lingered. IHC's long- and short-term debt went unserviced. Sales in all three groups dwindled after the Federal Reserve Bank raised an inflation-fed prime rate to 20 percent, and consumer rates flew higher still. In March, IHC realized that what labor demanded was less threatening than its financial crisis. Working at the bargaining tables almost without break, they settled on April 14, 1980, six months after the strike had begun. The strike had cost $579.4 million in losses during the first half of fiscal 1980.

No one at IHC could have predicted the future. Had the strike not occurred, would IHC still be in business? Maybe, but there were other pitfalls awaiting the corporation.

However, Engineering needed time with the Synchro-Torque system and the Vari-Range transmissions to perfect the compound planetaries, and Manufacturing was moving slowly due to the major tooling necessary. R. J. Roman, the TR-4 project manager, alerted J. T. Tracy, AEG's director of Product Planning and Development, that, "as a result, plans to incorporate these two new transmissions on 88-Series 2+2s would be delayed until May 1982."

While internal squabbles in late 1979 dealt with the products designated to restore IHC to first place, factors outside AEG were conspiring to make that much more difficult.

Hayford's 8760 philosophy came with very high overhead costs. In Hayford's perfect world, workers changed shifts in the instant between tasks when their replacement stepped in and picked up their tools. Real-world assemblers don't perform with the rehearsed precision of relay-team baton handoffs.

The market for all IHC's products had slowed. Yet Hayford saw only inefficient, idled plants that, in his view, cost the company money when they weren't turning out products that made the company money; that nobody was buying didn't matter. Fiscal 1980 ended with tractor sales 14 percent lower than 1979. Overall sales dropped by 29 percent. McCardell reduced his ambitious capital-spending plan for 1981 by one-third.

McCardell, despite his financial background, believed IHC's problems could all be solved if the company just had a couple of great selling seasons. Hayford agreed. Funding for the TR-4 and TR-3A programs continued.

In January 1980, 2+2 project manager Roman went back to AEG president Bill Warren to ask him for more money. A synchro-shift mechanism from Clark Equipment Company did not prove strong enough. A worldwide search for alternatives found nothing useful, so FEREC developed their own. It cost $8 million. Other component design and development consumed another $8 million, while manufacturing facilities intended for TR-4 use were not available. Farmall Works started to construct a plant addition.

The success of the 2+2s accelerated development of a 195-PTO-horsepower version. The drivetrain was marginal above 180 horsepower, so Engineering now considered the 195 model

as a segment of the Synchro-Torque TR-4 program that became the 7288 and 7488 2+2 "Super 70" models.

Meanwhile, failures during final testing of the 3788 tractor with the old 86-Series transmission suggested final-drive reliability problems would occur in the second or third year of average operation. The 3788 power train was a risk in terms of future farm tractor liability. These were not small problems. Engineers at the product reliability and support center near Hinsdale suggested these types of failures may require a rebuild of both the range and the speed transmissions at a cost of $3,300. The total: $5.5 million.

The 1985 Model 7288 2+2. Jeff Kelich put his 2+2 to work pulling his Model 720 6-18 On-Land plow through hard, dry soil in Indiana. The On-Land plow allowed the tractor to remain flat and level on the unplowed land to the left of the furrows. The 18-inch (45.7 cm) bottoms dug deep into the dry soil. Spring shock mounts kicked the plow bottoms out of the soil upon impact with a rock.

left: This was the operator's view from the 7288. Turning the wheel of the 2+2 swiveled the long nose ahead of the platform and brought the cab and rear axle around to follow it.

opposite: IHC produced the 7288 in 1985 only, and this is the fourth of just nineteen. Powered by IHC's DT-466 turbocharged diesel, the tractor boasted 175 PTO horsepower.

A frenzy of meetings followed in which Marketing worried over releasing a transmission that AEG knew would fail. Engineers fired back an analysis of its revisions and fixes. Instead of delaying it, AEG advanced the start of production to August 5, 1980.

Two months later, on October 7, Marketing approached Simmons from AEG's Business unit. Marketing was concerned about large 4WD sales. IHC's percentage of the business had fallen steadily since 1976, when it had reached 14.4 percent of the total market. By 1979 it was only 10.8 percent, and Marketing projected it at 8 percent for 1980. They feared that even with 1981 improvements to the Steiger articulated 4WDs, IHC still had no powershift or PTO option for those tractors, although they had been proposed. Simmons agreed.

In late October 1980, the TR-4 and TR-3A programs asked for another $9.4 million. Of this, $6.5 million would make up engineering time lost during the UAW strike. Now, AEG reassigned more than one hundred design engineers and managers who had shop skills to move the project forward. The remaining $2.9 million was for design, assembly, and testing of additional prototypes. Assembly of the TR-4 involved 220 new major machine tools requiring new automated and special transfer line equipment. A wary Manufacturing Department planned to build 100 preproduction tractors in May 1981.

In February 1981, Marketing's J. W. "Bud" Youle drove prototype TR-3A and TR-4 tractors at the Phoenix Proving Grounds. He was impressed and made that clear in his memo to Simmons and project manager Roman.

The situation inside IHC was approaching critical mass. In May 1981, to stanch the outward bleeding of cash, McCardell had "sacrificed" IHC's only profitable asset, the Solar Turbines International Group, to Caterpillar for $505 million. By early spring 1982, McCardell proposed yet another restructuring of management and debt. Hayford suggested taking the corporation into Chapter 11 bankruptcy. In March, fed up, Hayford resigned.

 The tractors came out: the TR-3A, 30-Series in August 1981, the 50-Series and revised 2+2s just reaching dealers. The economy that hamstrung IHC also hurt farmers. Federal bans on Soviet grain sales in 1980 gave foreign farmers a windfall market but left their own countries hungry. American farmers who no longer fed the communists nourished others briefly. But President Ronald Reagan's fiscal policies elevated the value of the dollar so greatly that few countries could afford American produce. By 1982 American farmers were becoming an endangered species. So were healthy farm equipment dealers.

Huge stocks of IHC's previous models remained on distributors' and dealers' lots. When the 30- and 50-Series arrived, churned out by plants running on the theory that large profits came from volume sales, they arrived at dealerships where few people were interested in, or capable of, buying them. Payment-In-Kind (PIK), a federal program to reduce crop acreages and surpluses, dealt farmers another severe blow when the 1983 drought reduced harvests in the fall. Foreign farmers hurried to feed Americans, accepting strong dollars in payment.

On August 10, 1983, Donald D. Lennox, the new president and chief executive officer replacing Hayford, signed off on

INTERNATIONAL
7488
IH

the Super 70s, the 7288 and 7488 models, with little additional development in response to Marketing's memo. Lennox and new AEG president J. D. Michaels released the Super 70s as well as the 175-horsepower 7288 model and the 195-horsepower 7488. Each of these offered full 50-Series features, including the Vari-Range transmission, a "Fuel-Efficient" engine management program, a 40-gallon (151 L) per minute high-pressure hydraulic pump, and the new Control Center cab.

The rear half of the 70-Series 2+2s was very similar to the rear of the 50-Series 2WD tractors, and it was manufactured on the same line. Shop staff then transferred that half out to a new 78,000-square-foot building completed in 1978, for $18.9 million ($68.6 million today). In this structure, dedicated to assembling and painting 2+2s, and manufacturing the front of the 70-Series tractors, workers joined the halves and completed final assembly.

Part of the launch program, scheduled for June 1984 through April 1985, included damage control. From October through December 1984, IHC held what it called customer and dealer Weather Vane meetings, "to improve the customer image of the 2+2 concept and to provide them with detailed changes as to why the new 2+2 will be more reliable in the future."

By January 1984, the feeling of concern spread far beyond North Michigan Avenue. Marketing wrote another memo late that month to AEG product marketing manager A. W. Williams, pointing out shortcomings of 50-Series tractors and Control Center cabs with door and control lever placements. Due to

drastically reduced capital spending and development budgets, corrections and improvements had to be slipped back to 1987 for introduction.

It was far too late. Lennox had launched a strategy to save the corporation in 1981. After its success that year, it gathered momentum. Tractor Equipment Company, a major parts supplier, heard that Dresser Industries, Inc., a Texas-based oil exploration and development conglomerate, was looking to add construction equipment to their line. IHC's Payline Group, with plenty of inventory on hand, fit their profile. Dresser struck a deal, acquiring the Hough operations as well as IHC Payline later that year. This left Lennox one less group to focus on. Through

INTERNATIONAL
5488

1984, he conferred with bankers and courted a potential buyer. On November 26, 1984, following hundreds of hours of work and negotiation, IHC agreed to sell AEG to Tenneco, Inc., of Houston, another large conglomerate with primary interests in oil production but with a wholly owned agricultural subsidiary, J. I. Case. IHC received $301 million in cash (about $686 million today) and $187 million in Tenneco preferred stock.

Case manufactured about twenty farm tractors. It had quit the harvester business long before, but it wanted IHC's axial-flow combines. Tenneco did not want IHC's Rock Island or Memphis plants. Overnight, Tenneco and IHC increased Case's market share to 35 percent, making it a contented runner-up to Deere, which had 40 percent. IHC brought thirty-three tractor models to the sale. The new company, Case IH, dropped the 2+2 and new 30- and 50-Series tractors because they didn't want Farmall Works, and they carefully selected the tooling they wished to own.

On May 14, 1985, the last International tractor, a Model 5488 All-Wheel Drive, came off the assembly line at Farmall Works. The Payline Group had moved to Texas, and Trucks remained in Chicago as Navistar. The Cub Cadet was safe with MTD Corporation in Cleveland, Ohio. But one of America's great legends was gone.

In 1983, due to an EPA requirement, Case had abandoned its Power Red and Power White paint scheme because there was too much lead chromate in its paints. It adopted a white-and-black combination until early in 1985, when sheet metal became Harvester red while the chassis remained Case black. Nearly all

of IHC's popular Neuss-built and Doncaster-produced tractors remained in production, providing Case IH much greater European name recognition than before. Case's own 94-Series tractors, built at its Racine, Wisconsin, plant, replaced all the domestic-built AEG models.

IHC's engineers had completed work on their powershift transmission. They mated this with Case chassis, engines, and bodywork. This combination, introduced by Case IH in 1988 and kept in production until late 1993, became the Magnum line of 130- to 195-PTO-horsepower two- and four-wheel-drive tractors.

above: The standard-equipment Synchro-Torque TR-4 transmission provided eighteen forward speeds with full synchromesh shift-on-the-go capability. With single rear tires, the tractor weighed 14,061 pounds (6,378 kg).

opposite: The 1984 Model 5488. These were among the last models to wear the International name. The company produced these in two- and four-wheel-drive versions.

above: The 1998 Case IH Magnum Model 8920. Case manufactured this series in two-wheel-drive and fitted with mechanical front-wheel-drive models. Production spanned 1997 and 1998. The six-cylinder turbo diesel displaced 505 cubic inches (8,275 cubic cm). The two-wheel-drive models weighed 15,630 pounds (7,090 kg). The mechanical front-wheel-drive hardware added another 945 pounds (429 kg). Both versions offered 155 horsepower at the PTO.

below: Case IH 1998 Maxxum Model MX-120. Case kept these tractors in production from 1996 into 2002. The company offered them with mechanical front-wheel drive, as shown here, as well as two-wheel drive.

opposite: Case IH 1998 Quad-Trac Model 9380. These were go-anywhere machines. Case manufactured this model from 1996 into 2001.

The 1990 Case IH Maxxum Model 7140. Case IH introduced these models in 1987, and they remained in production into 1994. The two-wheel-drive models appeared in 1988. Their Case IH six-cylinder 504.5-cubic-inch (8,267 cubic cm) turbo diesel developed 197.5 horsepower on the PTO. The four-wheel-drive models weighed 16,578 pounds (7,520 kg) with single rear wheels fitted. List price was $103,180, about $12,000 more than the two-wheel-drive version.

left: With twelve speeds forward and two available for reverse, these machines often found use on road construction sites. The big Quad-Trac weighed 43,000 pounds (19,504 kg). With its retail price set at $211,791, the Quad-Trac offered serious power and versatility for the operator who needed it. Its articulated steering and high drive sometimes allowed these machines to pull other crawlers out of the muck.

right: Case acquired engines from Cummins. This was the N-14 inline six-cylinder turbo diesel with 855 cubic inches (14,011 cubic cm) of displacement. Cummins and Case rated it at 355 horsepower at the PTO.

In September of that year, Case IH introduced the Second Generation Magnums.

Prior to acquiring IHC, Case engineers had begun work on the Maxxum Series, Case's World Tractor, offered with an adjustable front axle in 2WD or mechanical front-wheel-drive assist. IHC engineers joined Case early in that process. When Engineering completed its prototypes, they went to IHC's FEREC in Hinsdale for testing and customer marketing trials. After completing engineering work, the former IHC Neuss Works did the styling and assembly. Case IH introduced its 5100-Series in 1989 and the Second Generation 5200s arrived in late 1992.

In 1987 Tenneco had acquired Steiger Tractor Company of Fargo, North Dakota, which had previously produced articulated 4WD models for IHC. Steiger was a victim of the early

1980s farm crisis and had gone into Chapter 11 bankruptcy. Quickly rebadged, Case IH introduced its 9100-Series still using Steiger model names: Puma, Cougar, Tiger, and others in 1987. In August 1990, the Second Generation Case IH 9200-Series of Fargo-built 4WDs appeared. Nearly a year earlier, Steiger had begun field-testing in North and South Dakota a multitrack prototype based on the 9250. Tests went on carefully for three years, but few people saw the machine because operators ran it only in the dark where, from a distance, it was indistinguishable from the wheeled model. A shiny version toured Farm Progress and other major shows during 1992, but it was billed more as a "concept tractor" than a work-in-progress. Case IH meant to show Caterpillar that it was not alone in thinking about rubber tracks.

Nighttime tractor development continued, but Case IH also tested the technology on harvesters, especially in California's rice country north of Sacramento in late summer 1993. When they introduced the production Quad-Trac to dealers in Denver in 1996, those who'd seen the show model were surprised at the growth. It had gone from 246 horsepower to the 360-horsepower 9370 chassis. In mid-1998 Case IH introduced a second model, a 400-horsepower 9380. Model proliferation continued. The horsepower race never ended.

TEN

2000 *and Beyond*

THE RETURN OF FARMALL Robert N. Pripps

The 2014 Case IH Magnum 240 with a Case 1255 Early Riser Planter attachment. After 2010, several models, including the Magnum 240, featured Continuously Variable Transmission (CVT) that automatically selects the most efficient transmission range for the desired speed or load, eliminating the need for manual clutching and shifting. With a 6.7-liter six-cylinder diesel engine and a sticker price of $225,400, farmers paid top dollar for impressive power. *Courtesy CNH Industrial N.V.*

Before the takeover of CNH Global by Fiat in 1999, little was known in America about the giant Italian conglomerate. Since then, the company has become more well-known because of its acquisition of Chrysler Corporation. As a result of these two mergers, the complex Fiat Group has been in an almost constant state of flux, as it sorts out its overlapping brands and struggles to keep lines of communication open between its far-flung operations.

Fiat S.p.A. is the parent company of Fiat Group. In early 2011, Fiat Group separated automobile activity from Fiat Industrial S.p.A. Then, CNH Global N.V. and Fiat Industrial S.p.A. merged into CNH Industrial N.V., a company incorporated in November 2012 in the Netherlands. It became operational on September 29, 2013. It is listed as an American-Italian company that designs, produces, and sells agricultural and construction equipment, trucks, buses, and specialized vehicles, in addition to engines and power trains for industrial and marine applications. CNH Industrial is now the global leader in these markets.

At the time of the merger creating CNH Industrial N.V., CNH had more than 11,500 dealers spread across 170 countries. Manufacturing facilities were based in North America, Europe, Asia, and Latin America.

CNH Industrial N.V. is managed by a Group Executive Council (GEC) reporting to the company's board of directors. Sergio Marchionne chairs the Council. The GEC is composed of four main groupings: Agriculture, Construction, Commercial Trucks, and Iveco/Powertrain. Richard Tobin is the Group CEO and Andreas Klauser is COO for Europe, the Middle East, and Asia (EMEA) as well as brand president of Case IH Agriculture.

The Case IH brand and red logo embody a tradition of leadership in agricultural equipment. The brand is known for strong performance, low operating costs, and high reliability. Products include tractors, balers, and combines, each reflecting the heritage of such industry pioneering brands as McCormick, Deering, International Harvester, J. I. Case, and David Brown. Today, Case IH Agriculture is recognized for its powerful and highly productive agricultural equipment, and known also for its bright red livery. In 2004 Case IH Agriculture reintroduced the historic Farmall name and is currently expanding the line.

New Holland originated in 1895 and was acquired by Ford's Tractor Division in 1986. Then, in 1991, the New Holland Holding Company was formed. The assets of Ford New Holland and Fiat Agri were placed under its control. In 1993 Fiat completed the buyout of Ford's assets. Today, under the banner of CNH Industrial, New Holland offers more than one hundred product lines specifically for cash-crop producers, livestock farmers, vineyard owners, and ground care professionals.

Steyr has been a leading producer of tractors in Austria for more than sixty-five years. The trademark red-and-white design was created in 1967. Steyr's tractors are produced for the high-end buyer in Austria and are exported to Germany, Switzerland, Italy, Belgium, the Netherlands, and Luxembourg. In addition, there has been strong growth in sales to Poland, Hungary, Slovenia, and the Czech Republic. Nineteen different tractor models are offered.

Under Fiat ownership, the current Case IH tractor product lines took shape as early as 2004: 4WD (Steiger), Magnum, Maxxum, Puma, and Farmall. Each line has undergone more or less continuous development. Drawing from sister companies also under Fiat, new efficient engines and transmissions have become available for these tractors. Developments have always been with an eye on increased productivity. Despite the new ownership, care is being taken to ensure building on the traditions of Case and IH. This of course means continuing the rich red paint scheme that has been a trademark for more than seventy-five years.

4WD

At the top of the Red Power line are the huge articulated 4WD tractors. They come in row-crop (small-frame) or general purpose (heavy-duty) versions and with either wheels or, in the Quad-Trac crawler configuration, four rubber tracks replacing the wheels. An extra-heavy-duty frame is offered for industrial and construction use.

The STX Series of articulated 4WD tractors, built in Fargo, North Dakota, was continued from 2001 to 2007. The STX identifier was followed by a number indicating the horsepower: initially 275, 325, 375, and 440. After 2007, only the horsepower numbers were used and the "STX" was dropped. The Row-Crop version, using the smaller frame, could have 10-degree steerable front axles (as well as articulation steering). This gave much improved maneuverability. Normal accoutrements for these included a Category IV-N rear three-point hitch and a rear PTO with a 1,000 rpm option. Engines for the STX 275 were 8.3-liter units from CDC (Consolidated Diesel Corporation—a joint venture between Case and Cummins founded in 1980), while the STX 325 used a 9-liter Cummins. Both were fitted with sixteen-speed powershift transmissions. When ordered with AccuSteer, the steerable front axle, only the rear axle had a differential lock.

The STX 375-, 425-, 440-, and 450-Series used a heavier (larger) frame. In 2002 the 425 was added and the 450 replaced the 440. A 15-liter Cummins engine was used as was the sixteen-speed powershift transmission. The 425 and 450 versions were Tier 2 emissions–compliant. In 2003 a 24-volt starting

Case IH produced the Steiger STX275 from 2002 into 2005. This four-wheel-drive bear of a tractor generated 274 PTO horsepower and 212 drawbar horsepower. A wide-stance wheelbase measured 139 inches (353 cm) and the tractor, if you could figure out a way to weigh it, came in at 31,785 pounds (14,417 kg). *Courtesy CNH Industrial N.V.*

system was added. These heavy-frame tractors could be ordered as Heavy Duty (HD) scraper versions for pulling earthmovers. Besides an even heavier frame, these HD units were equipped with a heavy cable arrangement that tied the front and rear halves of the tractor together so that the center hinge did not have to take excessive loads when being assisted by a helper tractor. In 2005 an STX 500 model was added to the line with a 500-horsepower engine. It set a plowing record of turning 792 acres (295 ha) in twenty-four hours.

In 2006 Fiat modified the Steiger 4WD line, still using the STX identifier, but with new horsepower numbers: the STX 280, STX 330, STX 380, STX 430, STX 480, and STX 530. The top four models in terms of horsepower could be ordered in wheeled, Quad-Trac, or HD-Scraper versions. The STX 280 and 330 were small-frame row-crop tractors that used CDC engines of 8.3 and 8.5 liters, respectively. Interestingly, the STX 380 and 430 models used Fiat-Iveco engines of 12.9 liters. Iveco is another subsidiary of Fiat, specializing in commercial vehicles and ecological diesel and natural gas engines.

Only a year later, in 2007, the STX nomenclature was dropped altogether, and the tractors received another updating. Now, instead of the STX, the name "Steiger" graced the hoods. The Steiger Series consisted of models 335, 385, 435, 485, and 535. The 335 still used the 8.9-liter CDC engine. The 385 was the smallest version to be offered as a Quad-Trac. It was powered by the Iveco 12.9-liter inline six-cylinder engine. The 535 model still used the 15-liter Cummins engine. The Steiger 485 used

a unique Iveco turbo compound engine of 12.8 liters. In addition to the regular turbocharger that fed pressurized air to the intake manifold, a second "power-recovery" turbine added power directly to the crankshaft through a fluid coupling.

For 2010 revisions in the line were made to comply with Tier 4A emissions standards with Selective Catalytic Reduction (SCR) technology. This means the use of an exhaust fluid additive was required. Among other improvements for 2010 were a reshaped hood for better visibility and easier servicing.

The following table shows the details of the 2010 lineup.

MODEL	ENGINE	PTO HP	PEAK HP	RATED HP
STEIGER 350	8.7L TIER 4A	290	390	350
STEIGER 400	12.7L TIER 4A	340	446	400
STEIGER 450	12.7L TIER 4A	385	502	460
STEIGER 500	12.7L TIER 4A	430	557	500
STEIGER 550	12.7L TIER 4A	473	613	550
STEIGER 600	12.7L TIER 4A	473	669	600

2013–14 STEIGER 4WD TRACTORS

For 2013 Case IH again refreshed the 4WD line. All engines now were "in-house" from Case, Iveco, or FPT (Fiat Powertrain Technologies). All met Tier 4B, considered the final standards, emissions requirements.

All-new for 2013 was the Steiger Rowtrac, making available for the row-crop farmer the advantages of rubber tracks for increased traction and reduced soil compaction. For 2014 engine improvements were incorporated. The sixteen-speed powershift was continued throughout.

ROW-CROP CONVENTIONAL TRACTORS

Tractors from Case IH fall into one of two categories: articulated and conventional. The conventional tractors are sometimes called fixed-frame, or row-crop, and are characterized by larger rear wheels and smaller, steerable front wheels. The articulated are always 4WD; the conventional are sometimes rear-wheel drive only.

MODEL	ENGINE	PTO HP	PEAK HP	RATED HP
2013 STEIGER 350	8.7L TIER 4B	290	390	350
2013 STEIGER 400	12.7L TIER 4B	340	446	400
2013 STEIGER 450	12.7L TIER 4B	385	502	450
2014 STEIGER 370	8.7L TIER 4B	311	405	370
2014 STEIGER 420	12.9L TIER 4B	355	462	420
2014 STEIGER 470	12.9L TIER 4B	405	517	470
2014 STEIGER 500	12.9L TIER 4B	435	550	500
2014 STEIGER 540	12.9L TIER 4B	473	605	540
2014 STEIGER 580	12.9L TIER 4B	473	638	580
2014 STEIGER 620	12.9L TIER 4B	473	682	620

The 2014 Magnum 180 with L780 Loader. This 19,000-pound (8,618 kg) tractor could hoist, using an attached L780 Loader, 5,950 pounds (2,699 kg) to its full height. The Magnum 180's 6.7-liter FPT turbocharged intercooled diesel engine had six cylinders and measured 4.094 inches (10.4 cm) on the bore and 5.197 inches (13.2 cm) on the stroke. The tractor drew 180 rated engine horsepower and 133 horsepower at the PTO. *Courtesy CNH Industrial N.V.*

THE MAGNUM SERIES

Beginning life soon after the merging of Case and International Harvester, the 7100 Magnum Series was a new "from the ground up" tractor. It was first shown to Case IH dealers in August 1987 and went into production at that time. The first Magnum line consisted of four models: the 7110 (130 PTO HP), the 7120 (150 PTO HP), the 7130 (170 PTO HP), and the 7140 (195 PTO HP). Power for all was supplied by an 8.3-liter CDC. All were turbocharged, but only the 7140 employed an intercooler. A new deluxe cab was standard equipment; front-wheel assist was optional on all models. All models used an eighteen-speed powershift transmission, except the 7110, which got a twelve-speed unit, when not equipped with front-wheel assist. The central

Case IH manufactured the MX285 Magnum from 2003 into 2006. The 4WD tractor came standard with a turbocharged CDC 8.3-liter six-cylinder diesel engine and optional front-axle suspension that helped smooth the ride during long days in the field. This powerful tractor featured eighteen gears forward and four in reverse. *Courtesy CNH Industrial N.V.*

hydraulic system not only powered the three-point hitch and remote accessories, but also the steering, brakes, and transmission clutch packs. Production of this original Magnum Series continued through 1997.

For 1998 a new lineup of MX Magnum tractors emerged, which came to be known as Generation I. The line now consisted of five models with the following variations; all used the 18F-4R transmission.

MODEL	ENGINE	PTO HP	RATED HP
MX 180	CDC-T	145	180
MX 180 4WD	CDC-T	145	180
MX 200	CDC-TI	165	200
MX 200 4WD	CDC-TI	165	200
MX 220	CDC-TI	185	220
MX 220 4WD	CDC-TI	185	220
MX 240 4WD	CDC-TI	205	240
MX 270 4WD	CDC-TI	235	270

CDC=Consolidated Diesel Corp.
T=Turbocharger
TI=Turbocharger plus intercooler (same as aftercooler)

New styling for 1998 brought freshness to the cab and hood, plus the increased glass area gave improved operator visibility. Electronic fuel injection and draft control improved efficiency.

GENERATION II

Introduced in 2003, the second-generation MX Magnum Series again had four members now sharing a common platform (frame) and the 18F-4R transmission.

MODEL	ENGINE	TRANSMISSION	PTO HP	RATED HP
MX 210	CDC-TI	18F-4R	170	210
MX 210 4WD	CDC-TI	18F-4R	170	210
MX 230	CDC-TI	18F-4R	190	230
MX 230 4WD	CDC-TI	18F-4R	190	230
MX 255	CDC-TI	18F-4R	215	255
MX 285 4WD	CDC-TI	18F-4R	240	285

The standard version of the 210 and 230 came with 2WD; 4WD was optional. However, in 2004, the 2WD option was dropped due to low demand and the 4WD became standard equipment. Front-axle suspension was offered as an option on the MX 255 and the MX 285. This option greatly improved the ride.

While the engines have the same displacement, there are differences between them accounting for the different horsepower outputs. The MX 210 and 230 used a new type of Bosch mechanical fuel injection system, while the MX 255 and 285 used an electronic fuel injection system and had four valves per cylinder, rather than two. All of these engines were Tier 2 emissions–compliant (standards that took effect between 2001 and 2006).

Built from 2002 into 2007 at the Case IH plant in Basildon, England, the MXM 190 Maxxum was an impressive machine. Muscled to life by a turbocharged 7.5-liter six-cylinder diesel engine that drew 4.40 inches (11.2 cm) on the bore and 5.0 inches (12.7 cm) on the stroke, the tractor soared to 190 rated engine horsepower and 160 PTO horsepower. Four-wheel-drive was a tempting option, and full powershift provided nineteen forward and six reverse gears. *Courtesy CNH Industrial N.V.*

The 2006 MX305 Magnum. The year 2006 was the last for the MX name—in 2007 this tractor was sold as the Magnum 305. Driven by an 8.9-liter six-cylinder diesel engine that delivered 305 rated horsepower and 256 on the PTO, this 4WD tractor had a 118-inch (300 cm) wheelbase and weighed 21,585 pounds (9,791 kg). *Courtesy CNH Industrial N.V.*

GENERATION III

For 2006 Case IH again revised and refreshed the Magnum line. Since horsepower is the rate at which work is done, and there seems to never be too much horsepower when it comes to farm work, the new Series III Magnums had more horsepower across the board. New also was a 19F-4R powershift transmission providing a top speed of 25 mph (40.2 kph), used by all except the Magnum 275 with the FPT engine, which retained the 18F-4R transmission. In 2007 the MX designator was dropped, and the tractors were branded simply as Magnums, followed by the rated horsepower number. Also in 2007, the Magnum 335 was added to the line.

MODEL	ENGINE	PTO HP	RATED HP
MAGNUM 215 4WD	CDC 8.3L	178	215
MAGNUM 245 4WD	CDC 8.3L	204	230
MAGNUM 275 4WD	FPT 8.7L	248	275
MAGNUM 305 4WD	CDC 8.9L	256	305
MAGNUM 335 4WD	CDC 8.9L	277	335

GENERATION IV

For 2009 three new power levels of Magnum tractors, made by Steyr of Austria, a marque of CNH Industrial, were added to the lower end of the existing line. These tractors are similar to Puma models of the same number and were approved by the same University of Nebraska tests. All used the 19F-4R transmission except the Magnum 275 with the FPT engine, which used the 18F-4R transmission.

All in the line were now Tier 3 emissions–compliant and approved for biodiesel fuel. (Tier 3 standards were phased in between 2006 and 2008.)

MODEL	ENGINE	PTO HP	RATED HP
MAGNUM 180 4WD	FPT 6.7L	155	180
MAGNUM 190 4WD	FPT 6.7L	165	190
MAGNUM 210 4WD	FPT 6.7L	180	210
MAGNUM 215 4WD	CDC 8.3L	178	215
MAGNUM 245 4WD	CDC 8.3L	204	230
MAGNUM 275 4WD	FPT 8.7L	248	275
MAGNUM 305 4WD	CDC 8.9L	256	305
MAGNUM 335 4WD	CDC 8.9L	277	335

GENERATION V

Introduced in 2010 with Tier 4A emissions–compliant engines with Selective Catalytic Reduction (SCR), the Magnum line was again refreshed and improved. A new feature was the MultiControl armrest (also on Puma and Steiger models) that provided engine speed control, three-point hitch control, and transmission control at the operator's fingertips. All were now powered by FPT engines that offered Power Boost, a system that electronically allows as much as 35 more horsepower to be generated by the engine for road travel and/or mobile PTO or hydraulic loads without overtaxing the drivetrain.

Introduced in 2010, but expanded in application to more models through 2014, was the continuously variable transmission (CVT), a four-range gear differential device with automatic and seamless shifting between the four ranges. It provided a transport speed of 26 mph (41.8 kph). Only two ranges were available in reverse. There was a traditional "clutch" pedal that gave the operator the feel needed to hook up implements.

MODEL	ENGINE	PTO HP	POWER BOOST HP	RATED HP
MAGNUM 180	6.7L TIER 4A	150	234	180
MAGNUM 190	6.7L TIER 4A	165	250	190
MAGNUM 210	6.7L TIER 4A	180	260	210
MAGNUM 225	6.7L TIER 4A	195	269	225
MAGNUM 235	8.7L TIER 4A	195	274	235
MAGNUM 260	8.7L TIER 4A	215	298	260
MAGNUM 290	8.7L TIER 4A	240	328	290
MAGNUM 315	8.7L TIER 4A	265	358	315
MAGNUM 340	8.7L TIER 4A	290	389	340

For 2014, Tier 4B emissions standards were met and the CVT was available for all models.

MODEL	ENGINE	PTO HP	RATED HP
2014 MAGNUM 180	6.7L TIER 4B	155	180
2014 MAGNUM 200	6.7L TIER 4B	170	200
2014 MAGNUM 220	6.7L TIER 4B	185	220
2014 MAGNUM 240	6.7L TIER 4B	205	240
2014 MAGNUM 250	8.7L TIER 4B	205	250
2014 MAGNUM 280	8.7L TIER 4B	235	280
2014 MAGNUM 310	8.7L TIER 4B	265	310
2014 MAGNUM 340	8.7L TIER 4B	290	340
2014 MAGNUM 380	8.7L TIER 4B	315	380

New in the later part of 2014, the Magnum Rowtrac combined the best of wheels and tracks. Soil conditions can make getting into the field difficult for conventional wheel tractors. The Magnum Rowtrac "hybrid" could get onto soft ground without stalling or without excessive soil compaction. This adds up to earlier field work for higher yields and at reduced fuel consumption (due to less slippage).

MODEL	ENGINE	PTO HP	RATED HP
2014 MAGNUM	8.7L TIER 4B/FINAL	290	340
2014 MAGNUM	8.7L TIER 4B/FINAL	315	380

MAXXUM TRACTORS

As stated earlier, the Maxxum Series started life under Case before the merger with IH. There have been several iterations since its inception.

THE MXM MAXXUMS

These tractors, introduced in 2002, were based on New Holland designs and were built in Basildon, England. They use a Case six-cylinder engine of 7.5 liters equipped with a turbocharger and intercooler. Cabs were standard. All had mechanical 4WD and an 18F-6R powershift transmission. In 2005 lower-priced versions, MXM 130 and 140, were offered. These had conventional (mechanical) three-point hitches and manual transmissions. In 2006 the word "Pro" was added to the hood decal. These were equipped with heavier-duty front axles to better suit them for continuous front loader work and offered a partial-range powershift transmission. Initially, the company offered six models of Maxxum tractors, with the model number indicating the rated horsepower.

MODEL	ENGINE	PTO HP	RATED HP
MXM 120	CASE 7.5L	95	120
MXM 130	CASE 7.5L	105	130
MXM 140	CASE 7.5L	115	140
MXM 155	CASE 7.5L	125	155
MXM 175	CASE 7.5L	145	175
MXM 190	CASE 7.5L	160	190

THE MXU MAXXUMS

Built on New Holland platforms between 2003 and 2006, these tractors filled the gap between the MXM Maxxums and the smaller JX and JXU tractors. Cabs were optional in this line, but the cab offered was very fine indeed, with an air-ride seat and total soundproofing (72 dBA). Also featured was a roof window allowing the operator to see a fully raised loader.

Five model variations of the MXU line were offered. In 2006, the word "Limited" was added to the hood decal. All offered a twenty-four-speed shuttle powershift transmission option and a two-speed PTO.

MODEL	ENGINE	PTO HP	RATED HP
MXU 100	4-CYL., 4.5L	88	100
MXU 110	4-CYL., 4.5L, EFI*	102	110+15
MXU 115	6-CYL., 6.7L	105	115
MXU 125	6-CYL., 6.7L, EFI*	114	125+25
MXU 130	6-CYL., 6.7L	120	130

*Electronic Fuel Injection gave "Power Boost" to maintain engine speed in hard going.

The Maxxum line was again revised for 2012 with "Base" and "MultiController" series. They were much the same except the MultiController type had a six-function operator hand control (joystick). Those with the MultiController used the Semi Powershift or the CVT transmission. In the base model, without the MultiController, a Power Shuttle or a Semi Powershift were options. The styling was new, plus the hood was sculpted to allow for more front-wheel clearance for steering, reducing the minimum turning circle.

MODEL	ENGINE	PTO HP	RATED HP	POWER BOOST HP
MAXXUM 110	4.5L TIER 4A	90	110	143
MAXXUM 115	6.7L TIER 4A	95	116	154
MAXXUM 120	4.5L TIER 4A	100	121	154
MAXXUM 125	6.7L TIER 4A	105	125	165
MAXXUM 130	4.5L TIER 4A	110	130	163
MAXXUM 140	6.7L TIER 4A	120	140	176

JX AND JXU SERIES

When Case IH merged with New Holland in 1999, the European antitrust regulators forced the sale of the low-priced C and CX lines of tractors to Landini, leaving a gap in the product line. In 2002 CNH Industrial filled this gap with the JX and JXU Series.

The JX line was built in Ankara, Turkey, providing five variations of lower-priced utility tractors of from 42 to 80 PTO horsepower. The three- and four-cylinder diesel engines were built by Iveco with only the 80-horsepower unit being turbocharged. All were available with cabs and two transmission choices: a 12f/12r shuttle or a 20f/12r with creeper gears.

The JXU Maxima (JX Upgrade) line of tractors, made in Jesi, Italy, consisted of three- or four-cylinder JX tractors with

improvements. The same Iveco engine was used, but horsepower was marginally increased. Transmission choices were now a 24f/12r powershift or a 20f/12r with creeper gears. As with the JX tractors, front-wheel assist was standard and cabs were optional.

THE PUMA SERIES

Puma was introduced as a new series of row-crop tractors in 2007. The name had previously been used by Steiger for their smallest articulated tractor line. The new Puma Series was designed to fill in between the Magnum and Maxxum lines and was initially offered in short and long wheelbase versions. After 2008, all versions used the 113.6-inch (288.5 cm) wheelbase. The Puma was designed to be efficient in multitask farm operations with big-tractor features and small-tractor economies on lighter jobs. A 6.7-liter six-cylinder engine, turbocharged and intercooled, was used throughout. An eighteen-speed powershift transmission was standard (with six in reverse), but a nineteen-speed unit was an option that gave 31 mph (50 kph) transport speed. The Steyr-developed CVT was also offered after 2008. It too offered a top speed of 31 mph (50 kph).

The 2007–09 Pumas came in the following model numbers indicating rated horsepower.

MODEL	ENGINE	PTO HP	RATED HP
PUMA 115	6 CYL., 6.7L	95	115
PUMA 125	6 CYL., 6.7L	105	125
PUMA 140	6 CYL., 6.7L	120	140
PUMA 155	6 CYL., 6.7L	135	155
PUMA 165	6 CYL., 6.7L	135	165
PUMA 180	6 CYL., 6.7L	150	180
PUMA 195	6 CYL., 6.7L	165	195
PUMA 210	6 CYL., 6.7L	180	210

The Puma line was equipped for operator comfort with a modern soundproofed cab and front axle suspension. A category II/II-N three-point hitch (front and rear options) with electronic draft control in the rear was provided. The two-speed continuous PTO (540/1000 rpm) was reversible. At first, the fuel tank held 87 gallons (329.3 L), but later, that was increased to 107 gallons (405 L). Later engines are compatible with biodiesel fuel.

2010 MODEL PUMA TRACTORS

MODEL	ENGINE	PTO HP	RATED HP
PUMA 130	6.7L TIER 4A	105	130
PUMA 145	6.7L TIER 4A	120	145
PUMA 160	6.7L TIER 4A	135	160
PUMA 170	6.7L TIER 4A	140	170
PUMA 185	6.7L TIER 4A	150	185
PUMA 200	6.7L TIER 4A	165	200
PUMA 215	6.7L TIER 4A	180	215
PUMA 230	6.7L TIER 4A	195	230

2014 MODEL PUMA TRACTORS

MODEL	ENGINE	PTO HP	RATED HP
PUMA 150	6.7L TIER 4B	125	150
PUMA 165	6.7L TIER 4B	140	165
PUMA 185	6.7L TIER 4B	155	180
PUMA 200	6.7L TIER 4B	170	200
PUMA 220	6.7L TIER 4B	185	220
PUMA 240	6.7L TIER 4B	205	240

L745
CASE IH
200A

THE NEW FARMALLS

In 2004 Fiat reintroduced the Farmall name, rightly realizing the inherent goodwill and loyalty that went with the marque. The Farmall reputation for durability and reliability led to an enviable reputation for productivity. The "New Farmalls" capitalized on that reputation and proudly continued the red tractor tradition. Farmalls now come in three configurations: Utility, Specialty, and Compact. The current line was introduced in 2012 after several years of Farmall development. CNH reentered the compact tractor market in 2001 with the D and DX Series based on the New Holland Boomer line. These morphed into the D-DX and JX-JXU in 2008, and finally, into the Farmall Series, brought out in 2012 and updated in 2013 and 2014.

UTILITY FARMALLS

Farmall 100A Series

This series, made in Mexico, comes with four-cylinder diesels (4.5 liters) in the 110A and 120A versions, and with six-cylinder diesels (6.7 liters) in models 125A and 140A. All engines are from FPT and are turbocharged and have intercoolers. Options include ROPS or cab; two-wheel drive or mechanical front-wheel assist; and an 8x8 mechanical, a powershift shuttle, or a 16x8 powershift shuttle.

MODEL	ENGINE	PTO HP	RATED HP
FARMALL 110A	4 CYL., 4.5L	90	110
FARMALL 120A	4 CYL., 4.5L	96	118
FARMALL 125A	6 CYL., 6.7L	105	125
FARMALL 140A	6 CYL., 6.7L	115	140

Farmall U Series

These are heavy-duty utility tractors, made in Italy, with features aimed at livestock and haying operations and for heavy loader work. Engines, by FPT, are four-cylinder turbocharged with intercoolers, 4.5 liters for the 105 model, and 3.4 liters for the 115 model, according to the University of Nebraska tests. Options include a ROPS or a cab, two-wheel drive or mechanical front-wheel assist, and a 12x12 or a 24x24 powershift shuttle.

MODEL	ENGINE	PTO HP	RATED HP
FARMALL 105U	4 CYL., 4.5L TIER 4A	90	105
FARMALL 115U	4 CYL., 3.4L TIER 4A	98	115

The 2013 Farmall 105C with the factory-installed L630 Loader. The tractor's four-cylinder, 3.4-liter engine provides 106 rated horsepower and 91 on the PTO. As well as going back and forth, this tractor goes up forcefully, with a lift capacity for a 105C with the self-leveling option at 3,428 pounds (1,555 kg). Maximum lift height is 135 inches (343 cm), a little more than 11 feet (3.4 m). *Courtesy CNH Industrial N.V.*

opposite: Built in Jesi, Italy, from 2008 into 2012, the Farmall 75C was the smallest Utility Farmall of the C-Series, powered by a 3.2-liter four-cylinder diesel Fiat Powertrain Technologies (FPT) engine. The 75C engine produced 74 rated horsepower, while it came in at 65 horsepower on the PTO. Somewhat surprisingly, you could attach a front-end loader, and 4WD and a cab were other options that added to the sticker price. Weight, depending on the options you chose, could rise to 6,768 pounds (3.070 kg). *Courtesy CNH Industrial N.V.*

Farmall C Series

The C Series tractors are smaller and lower priced than the U Series, but feature power, comfort, and convenience that make them ideal for livestock operations and heavy loader work. These tractors have FPT diesels, are turbocharged and intercooled, and are made in Italy. Options include a ROPS or a cab; two-wheel drive or mechanical front-wheel assist; and an 8x8, a 12x12, or a 24x24 powershift shuttle.

MODEL	ENGINE	PTO HP	RATED HP
FARMALL 75C	4 CYL., 3.2L TIER 4B	65	74
FARMALL 85C	4 CYL., 3.4L TIER 4A	70	84
FARMALL 95C	4 CYL., 3.4L TIER 4A	82	98
FARMALL 105C	4 CYL., 3.4L TIER 4A	91	106
FARMALL 115C	4 CYL., 3.4L TIER 4A	98	115

Farmall A Series

Series A tractors are rugged, reliable tractors with 8x8 mechanical shuttle transmissions that are fuel efficient and easy to operate. Cabs are optional, as is mechanical front-wheel assist. They offer ample hydraulic power for blade and loader work. The A Series made its debut in 2012.

MODEL	ENGINE	PTO HP	RATED HP
FARMALL 45A	SHIBAURA 4 CYL., 2.2L	39	45
FARMALL 55A	FPT 3 CYL., 2.23L T	47	53
FARMALL 65A	FPT 4 CYL., 3.2L T	57	65
FARMALL 75A	FPT 4 CYL., 3.2L T	66	75

SPECIALTY FARMALLS

Farmall V tractors are designed for vineyard rows; Farmall N tractors are narrow and low and are designed for applications where space is limited. Options include cabs, front-wheel assist, and shuttle shifts. The Specialty tractors were not tested at the University of Nebraska.

MODEL	ENGINE	PTO HP	RATED HP
FARMALL 75N	4 CYL., 3.2L T	62	76
FARMALL 95N	4 CYL., 4.5L T	82	95
FARMALL 105N	4 CYL., 4.5L T	92	106
FARMALL 105V	4 CYL., 4.5L T	92	106

COMPACT FARMALLS

With deluxe cab, CVT, mechanical front-wheel drive (MFD), and shuttle shift as standard equipment, these are serious, hardworking professional machines—with power more reminiscent of the Farmall M than the original "compact" Farmall B of the 1940s.

MODEL	ENGINE	PTO HP	RATED HP
FARMALL 40B CVT	4 CYL., 2L	32	40
FARMALL 45B CVT	4 CYL., 2.2L	36	45
FARMALL 50B CVT	4 CYL., 2.2L	40	50

Farmall C Series

These are compact tractors with big tractor features. A large operator platform with a comfortable seat makes long workdays possible. The 30 and 35 models use a three-cylinder diesel of 1.5 liters, turbocharged; the 40 and 50 models use a 2.2-liter four-cylinder diesel, naturally aspirated. Transmission options mechanical shuttle shifts or a three-range hydrostatic transmission.

MODEL	ENGINE	PTO HP	RATED HP
FARMALL 30C	3 CYL., 1.5L	26	32
FARMALL 35C	3 CYL., 1.5L	29	36
FARMALL 40C	4 CYL., 2.2L	32	40
FARMALL 50C	4 CYL., 2.2L	37	46

Back in 1902, when George W. Perkins, a J. P. Morgan partner and adviser to the McCormick family, picked the name *International Harvester*, he had no way of knowing just how prescient he was being. Harvester, as it came to be known, was indeed an international company virtually from its beginning. Harvester's integrated team of experienced designers and marketers, backed by Morgan's financial intuition, not only dominated the U.S. and Canadian agricultural equipment markets, but also continually expanded into Europe, South America, and finally worldwide.

When Tenneco joined Harvester with Case to form Case IH, the company's reach expanded exponentially. Then New Holland was absorbed, forming CNH Global; and finally, now under Italy's Fiat conglomerate, it is truly an international company, although still with its American vestiges. This book is a celebration of the history of the great red tractors and how they revolutionized farming from mere backbreaking subsistence to its honorable career status enjoyed today.

But the agricultural industry still faces daunting challenges. Today, farmers feed seven billion people. By the middle of this century, 2050, that number is expected to exceed nine billion, a 35 percent increase. However, as more of the population advances in prosperity, the demand for protein (meat) will increase, requiring a doubling of farm output. The great red tractors, and others like it, allowed a reduction in farm labor from 18 percent of the U.S. population at the time of the Farmall's introduction to a mere 1 percent today.

The challenge for the twenty-first-century farmer is not so much to reduce labor as to increase production while minimizing the increases in land, fuel, and water usage and limiting atmospheric pollution. To meet the food needs of 2050, yields must be increased while dramatically reducing the environmental impacts of traditional farm practices. The future, in large part, depends not only on the farmer but also on agricultural universities and major commercial agricultural enterprises, including the tractor and harvesting machinery companies. Farmall tractors and other machines built by Fiat Industrial will play a significant role in meeting the world's expanding food needs.

The Farmall 35C is a little tractor with some descent muscle, including a 1.5-liter three-cylinder engine with 36 horsepower on the drawbar and 29 horsepower on the PTO. It had 1,807-pound (820 kg) lift capacity and a maximum forward speed of 14.2 mph (22.9 kph). Its wheelbase of 66 inches (167.6 cm) would make it seem that the 35C could drive right under the wheels of some other Case IH tractors, but what would be the point of that? *Courtesy CNH Industrial N.V.*

Brimming with creative inspiration, how-to projects, and useful information to enrich your everyday life, Quarto.com is a favorite destination for those pursuing their interests and passions.

Inspiring | Educating | Creating | Entertaining

title page: At Keith Feldman's farm, it was a gathering of the 40-Series Row-Crop family. The 240 is at left, the 340 is in the rear, and the 140 is in the foreground.

contents page: The 1919 International 8-16. The 8-16s were IHC's first products manufactured on assembly lines. Both engines and tractors advanced along a moving conveyor during assembly.

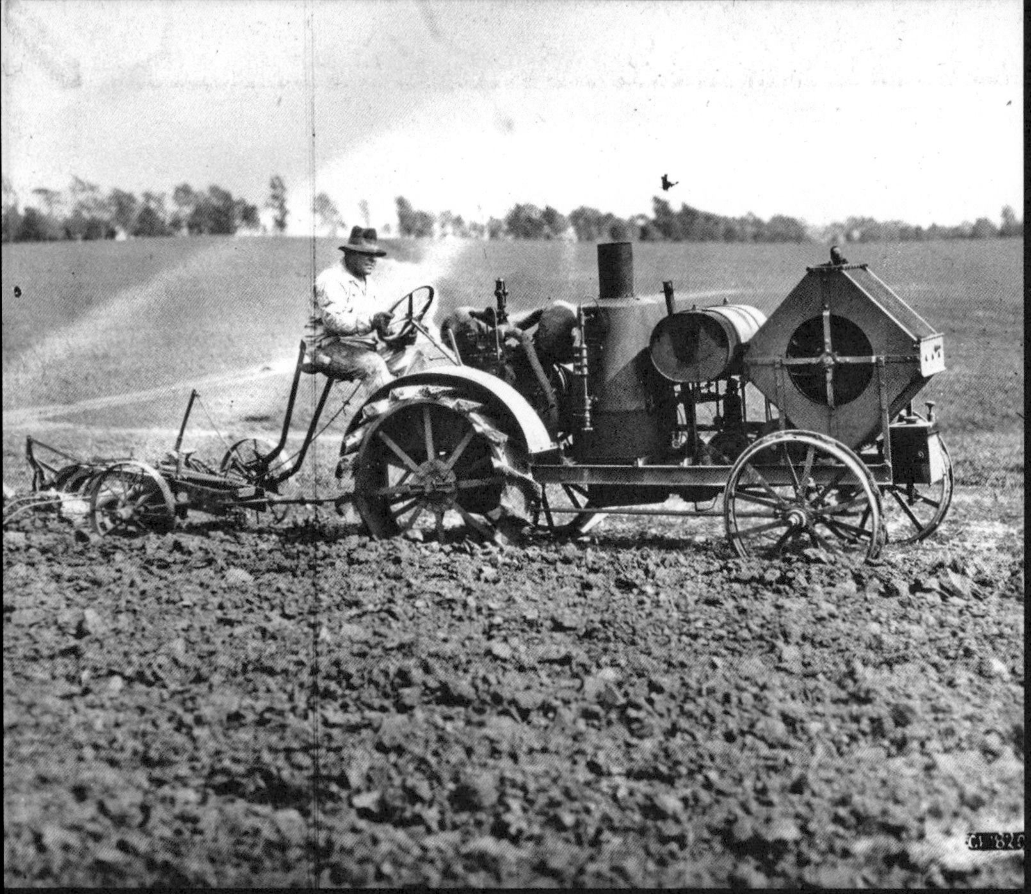

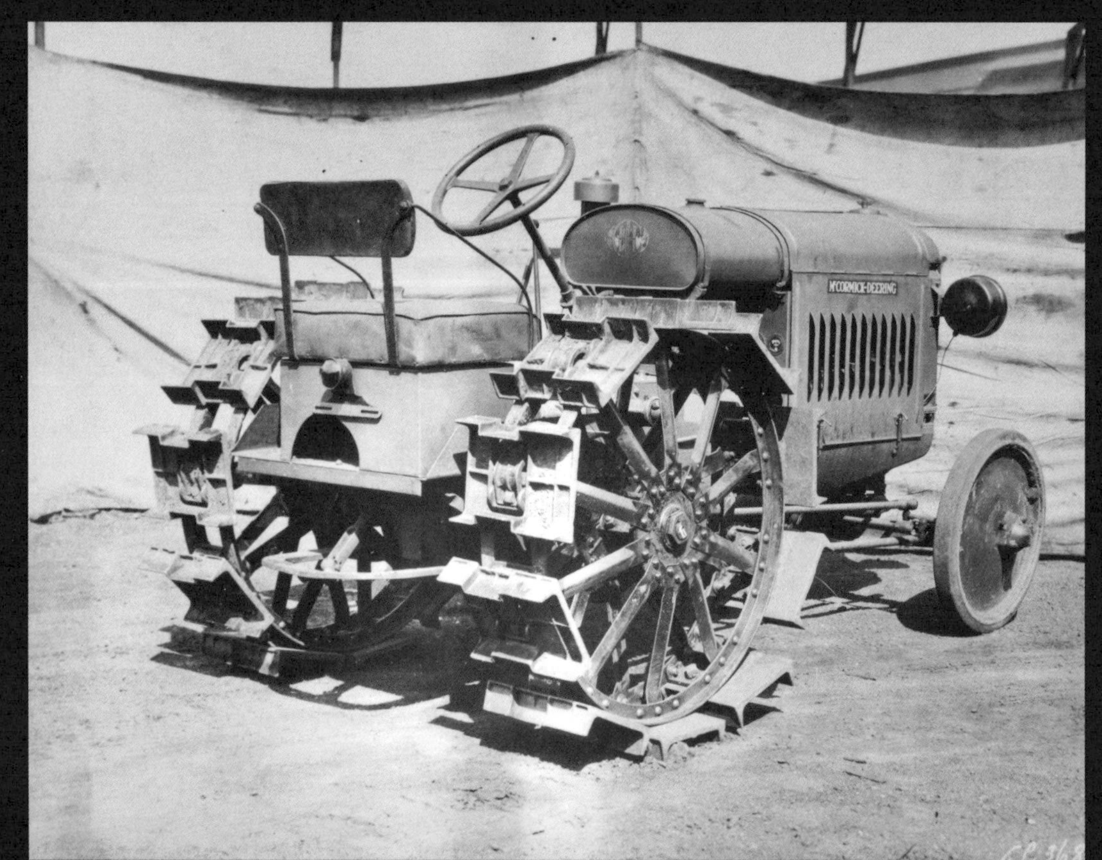

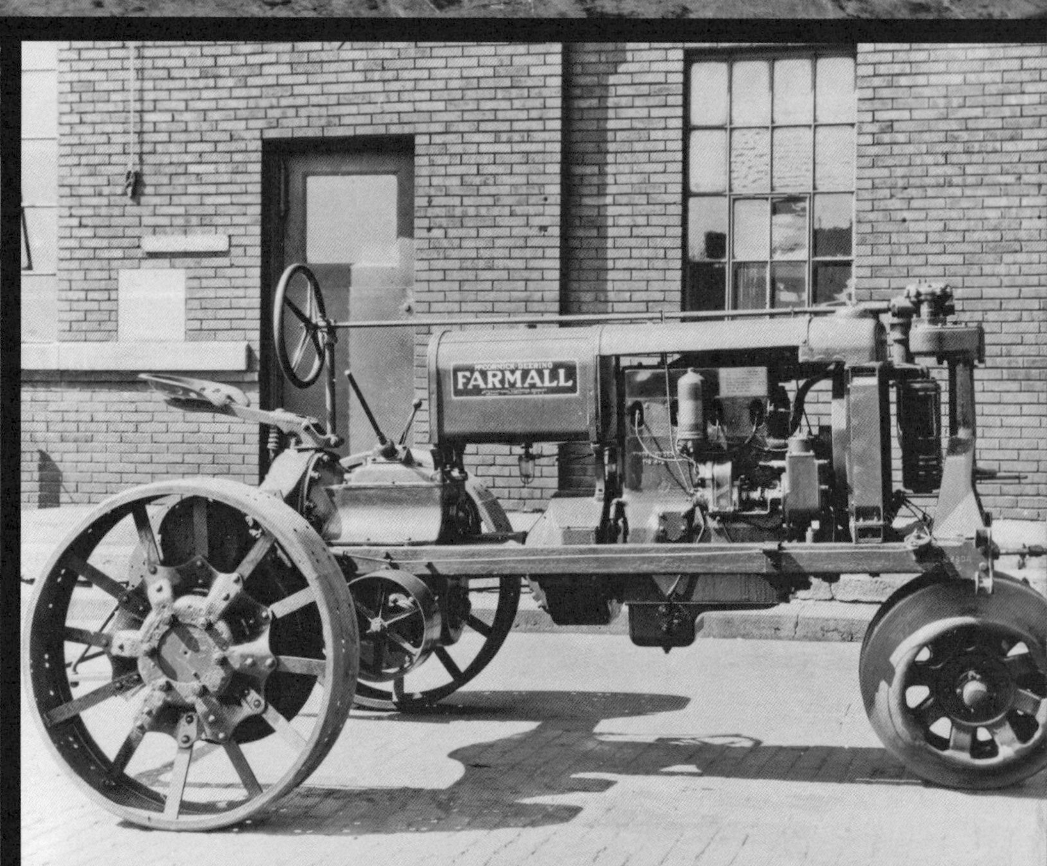

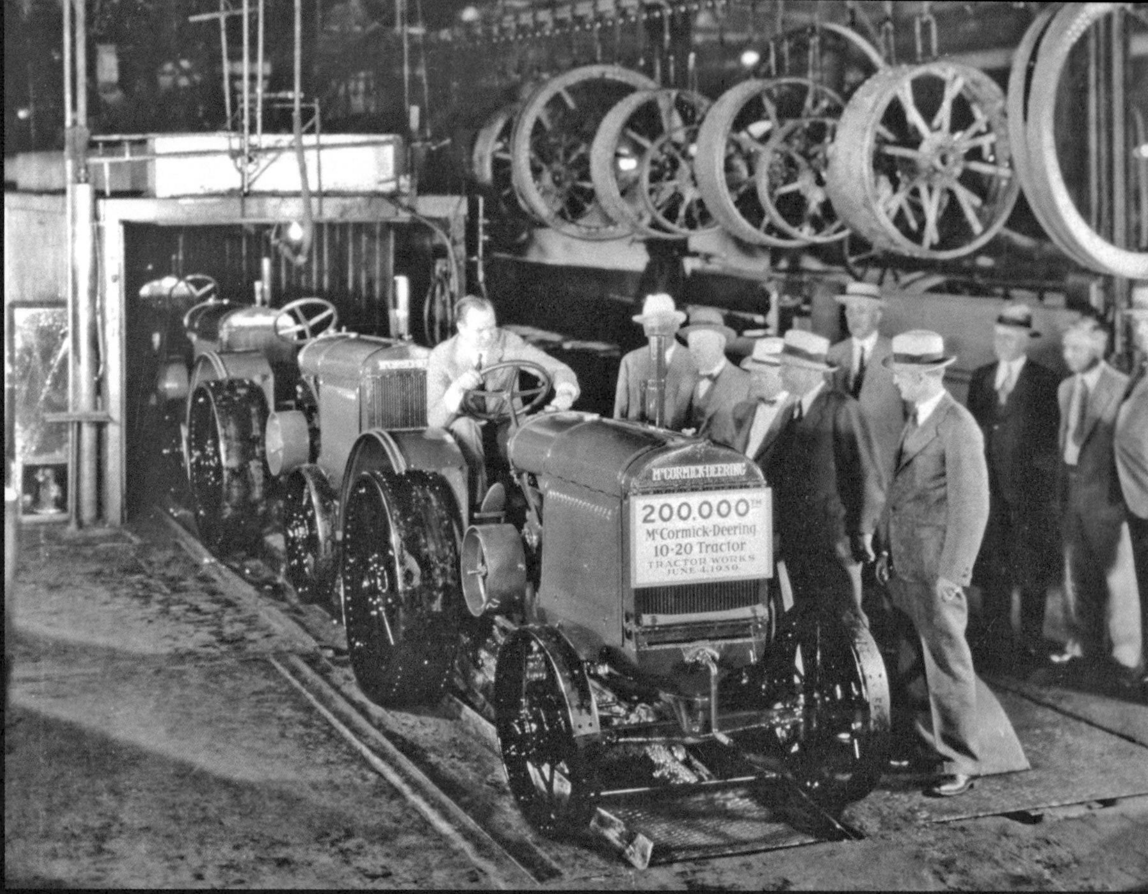
McCORMICK-DEERING
200,000th
McCormick-Deering
10-20 Tractor
TRACTOR WORKS
JUNE 4, 1930

McCORMICK-DEERING

McCORMICK-DEERING
FARMALL